138°

142°

PACIFIC OCEAN

BIAK

Biak

PEN.

Jayapura

IRIAN

JAYA

ORANGE MTS

SSAU
ISSEL
AKES

MTS.

Ertsberg

Kokonao

Amamapare

Omowka

PAPUA

NEW

GUINEA

Map of
Overland Routes
See Back Endpaper

SEA

AUSTRALIA

© 1981 A·Karl/J·Kemp

THE CONQUEST OF
COPPER MOUNTAIN

FORBES WILSON

THE CONQUEST OF
COPPER
MOUNTAIN

New York **ATHENEUM** 1981

Library of Congress Cataloging in Publication Data

Wilson, Forbes.
    Copper mountain.

    1. Copper ores—Indonesia—Irian Jaya. 2. Copper
mines and mining—Indonesia—Irian Jaya. I. Title.
TN446.I52I748   1948   338.2'743'09951   80-69396
ISBN 0-689-11153-3

ACKNOWLEDGMENTS

This is a very personal book about the most important single professional accomplishment of my life. It is my account, extending over 20 years, of the discovery, exploration, and development of the largest and richest copper deposit ever found above ground, a mountain of ore which is located in perhaps the most remote, primitive, and inhospitable area of the world. While I am proud of my role in making the venture a commercial success, it was small compared to the enormous efforts of many hundreds of other people who persevered against and overcame what often seemed like hopeless odds. After the backbreaking two-year task of building a sixty-three-mile access road through swamps and jungles and up steep mountain ridges, one tough veteran of many construction projects came up to me and said, "Mr. Wilson, when I first came here two years ago, I didn't think there was any way we could break through these mountains. But, by God, we have done it." It was that kind of collective spirit that made it possible for us to prevail.

For the record, I would like to set forth the names of the men with whom I was most proud to be associated from the inception of the Ertsberg project in 1959 to the completion of production facilities in late 1972, when I ceased having direct responsibility for and day-to-day contact with the project. Though I take responsibility for everything I have written, this is their book as much as it is mine.

THE PIONEERS: Jean Jacques Dozy, Jan van Gruisen, Del Flint, Gus Wintraeken, Jan Ruygrok, Moses Kelangin.

THE EARLY DEVELOPERS: Balfour Darnell, Del Flint again, Reg Barden, and John Currie.

THE NEGOTIATORS: Bob Hills, Julius Tahija, Ali Budiardjo, Henry Pierson, Bob Duke, Nils Kindwall, Hyde Brownell, Pete Petersen, Bill Byrne, Dr. Anondo, Saleh Bratanata, Soetarjo Sigit, Ed Rutledge of the Export-Import Bank, and Ambassador Marshall Green, who provided valuable advice and encouragement.

THE ENGINEERS AND CONTRACTORS: Steve Bechtel and many others from his company, and the following from Freeport: Willie Williams, Ed McNamara, Bob Wernet, Sid Chalmers, Steele Ansley, Orrin Main, and Buck Steele.

THE SMELTERS: Eddie DeCouto of Metal Traders and Shimpei Omoto of Mitsui Mining Company in Japan, Walter Gleich of Norddeutsche Affinerie and Erich Bachem of Kreditanstalt für Wiederaufbau in Germany.

THE OPERATING GROUP: First, Usman Pamuntjak, an Indonesian engineer who joined our organization during the early days of construction and has moved up in the organization to his present position of Manager of Operations. Second, to the General Managers each of whom in their time (1972–1980) has faced and overcome unusual and difficult problems: Alan Latham, Dick West, Morgan

Mickleberry, Bob Weaver and Les Acton. And last, but certainly not least, Milt Ward, who succeeded me as chief operating officer, and who for the past six years has provided the cement which has kept the Ertsberg operating structure together.

I want to give special thanks to Paul Douglas for his encouragement. I had written up my notes from the 1960 overland expedition during the 1962–64 hiatus, when the project was moribund due to political developments in Indonesia. When the project got underway again in 1965, I put them aside. In 1974, after my retirement, Paul urged me to complete the manuscript and said Freeport would be willing to assist with its publication. For post-1972 developments, I have relied heavily on information and observations furnished to me by others at Freeport.

I would like to acknowledge, as well, the work of business writer Chris Welles, who expanded the manuscript and put it into final publishable form.

And finally, a word for my wife, Ann, who endured my long absences from the family circle, who listened with patience and understanding to lengthy accounts of trials and tribulations I brought home from faraway places, and who, after 14 years of wondering what it was all about, was rewarded in 1974 with a visit to the Ertsberg and a trip around the world.

Forbes Wilson
December 30, 1980

Pg. 129

CONTENTS

ILLUSTRATIONS

xi

Barge stands at port to receive copper concentrates from mine.
(FREEPORT MINERALS COMPANY)

Foundations under construction for tramways, 1972. (FREEPORT MINERALS COMPANY)

Tram car carrying ore chunks down the mountain. (FREEPORT MINERALS COMPANY)

The car releases its load. (FREEPORT MINERALS COMPANY)

Pipeline laid along trench that was cut through coastal jungles.
(FREEPORT MINERALS COMPANY)

Aerial view of mining operation at the Ertsberg. (FREEPORT MINERALS COMPANY)

Freeport-built town of Tembagapura. (CHRIS WELLES)

Workers commute to mill and mine sites. (CHRIS WELLES)

Helicopter-drops supplement Tembagapura's food supply. (CHRIS WELLES)

Indonesian miners descend road to open-pit mine. (CHRIS WELLES)

Mining crew at the Ertsberg East deposit. (FREEPORT MINERALS COMPANY)

Villagers, near Tembagapura, in tribal dress. (CHRIS WELLES)

Local inhabitants trained as mechanic's helpers. (CHRIS WELLES)

Nurse at Tembagapura hospital treats foot injury. (FREEPORT MINERALS COMPANY)

Forbes Wilson discussing engineering problems, 1972. (FREEPORT MINERALS COMPANY)

Forbes Wilson and geologist Frank Nelson, 1976. (FREEPORT MINERALS COMPANY)

THE CONQUEST OF
COPPER MOUNTAIN

Prologue

I awoke just after dawn. Lying on my cot looking through the open end of the tent, I was immediately conscious of a gleaming white shape far in the distance. The sight was so strange that for a moment I could not imagine where I was or what I was looking at. Then my head began to make order of what my eyes had focused on. The white shape was a large patch of ice—actually a glacier—covering the top of the most forbidding looking mountain I had ever seen. Between our camp and the mountain were about 12 miles of green jungle, still cloaked in shadow. Above the jungle, at a height of 16,500 feet, the ice seemed to have been ignited by the first rays of the sun.

What made the view of the glacier particularly incongruous was that we were only a few degrees south of the equator. Our camp was at 6,000 feet, but just ten days earlier I had been at sea level, sweltering in the oppressive heat of the tropics. During those ten days, our expedition had battled its way 60 miles through some of the most treacherous terrain in the world: almost impassable coastal mangrove swamps, dense rain-forest jungles, cliffs that soared vertically for thousands of feet. There had been no established trail for most of the way. Often through thick fog and torrential downpours, we had followed the banks of swiftly flowing rivers, climbed through waterfalls, walked along jagged knife-edge ridges. In many places we had to pull ourselves up on jungle roots and vines.

The glacier stretches nearly four miles along what was then called the Carstensz Top, a peak named after Jan Carstensz, a Dutch navigator who, in 1623, on what must have been an exceptionally clear day, spotted it from the sea. It is the highest point on the great chain of almost unbelievably rugged mountains that forms the East-West backbone of what traditionally was known as New Guinea, the second largest island in the world, which lies between Australia and Asia. The Carstensz Top is situated on the western half of the island, a former Dutch colony then called Netherlands New Guinea. Western New Guinea became a province of Indonesia in 1963 and is now called Irian Jaya.* With 162,900 square miles, it is roughly the size of California. The eastern half of New Guinea, a for-

---

* Indonesian president Suharto renamed the province in 1973. Between 1962 and 1973, when administration of the area was first transferred from Holland to Indonesia, it had been called Irian Barat or West Irian. The names of most of Irian Jaya's other features have now been Indonesianized. The Carstensz range, which had been recorded on maps as the Snowy Mountains and later the Nassau Mountains, are now the Jayawijaya Mountains. The Carstensz Top is called Puncak Jayawijaya. With very few exceptions—mainly discussions of present-day conditions in the last chapter—this book uses pre-Indonesianization names.

mer Australian colony that won independence in 1975, is called Papua New Guinea.

The central highlands of Western New Guinea have been less explored than any other area of the globe, with the possible exception of some parts of Antarctica. Less is known about some of its geography than about much of the surface of the moon. Aerial relief maps are still sketchy and contain large sections labeled "Relief Data Incomplete." Until recently, the altitude of the Carstensz Top was recorded on many maps as 11,000 feet, a 5,500-foot error that has had fatal consequences. The crumpled remains of several airplanes can still be seen on the peak's upper slopes.

Nearly all of the lowlands of Western New Guinea are covered by swamps and jungles. One popular guidebook, *Indonesian Handbook* by Bill Dalton, calls the jungle "more impenetrable and treacherous . . . than any other tropical region, including the Amazon Valley and Africa." Ross Garnaut and Chris Manning, the Australian authors of *Irian Jaya,* comment: "Few parts of the earth's surface constrain man's movements as effectively as the swamps, jungles, and mountains of New Guinea."

Western New Guinea is so inhospitable that it is inhabited only sparsely by perhaps a million primitive Melanesians once known as Papuans and now called Irianese, who are divided into about 150 tribes. The tribes have had so little contact with each other that they speak mutually unintelligible languages and dialects. Regarded by educated Javanese in the western part of the country as the "Siberia" of Indonesia, the island has been almost totally cut off from the modern world. Until the last few decades, the only outsiders to live there were a few hardy Western missionaries in settlements along the coast. Some tribes in the mountains have never seen a white man and live a virtual Stone Age existence. A few still practice head-hunting and cannibalism. Four Dutch families were killed and eaten by the inhabitants of a mountain village as recently as 1974.*

As excited as I was by my first view of the Carstensz Top, my mission was not mountain climbing. I was the manager of minerals exploration for a large American min-ing company then known as Freeport Sulphur and now called Freeport Minerals Company, with sales in 1980 of $664 million. My reason for being in New Guinea was not the Carstensz Top at all, but a less lofty and osten-sibly less impressive hill nearby, a dome-shaped black mass at an elevation of nearly 12,000 feet that rose up more' than 500 feet from the floor of a narrow valley. A young Dutch petroleum geologist named Jean Jacques Dozy, who discovered it in 1936, named the outcrop the "Ertsberg," which is Dutch for "ore mountain." (In Indo-nesian, it is known as "Gunung Bijih.") My expedition had landed on the coast in early May of 1960 to find out just how much ore the Ertsberg contained.

I have had a lifelong interest in geology since my gradu-

---

* Another purported victim of cannibalism was Nelson Rockefeller's son Michael. In 1961, while he was on an expedition to the south coast of New Guinea to obtain native artifacts, the boat in which he and a friend were sailing capsized. The friend was later rescued, but Michael, who tried to swim the 11 miles to shore, was never seen again. There were many rumors that he had been eaten by natives. My own view is that he never made it through the shark- and crocodile-infested coastal waters.

ation from Yale in 1931 and have done exploration work in many parts of the world, but my previous experience with roughing it in the mountains was limited to riding mules through the Andes during the 1930's prospecting for gold. I had never undertaken any overland hikes even remotely resembling the one I was on in New Guinea. It had been far more strenuous than I had ever anticipated.

Jean Jacques Dozy had no inkling of it in 1936, but the Ertsberg later turned out to be the largest known above-ground copper deposit, 33 million tons of high-grade ore worth over a billion dollars at 1980 prices. Below-ground copper deposits were recently discovered nearby that ultimately may turn out to have five or more times as much copper as the Ertsberg.*

Despite the Ertsberg's riches, I'm still not sure I would

---

* The first metal known to man, copper came into wide use during the Bronze Age, 3,500 to 1,000 B.C. The name derives from the Mediterranean island of Cyprus, which was the chief source of the metal in the ancient world. The Romans called it *aes Cyprium* or "Cyprium metal," which was corrupted to cuprum and ultimately to the English "copper."

Worldwide production is about 8 million tons annually, and copper prices are highly cyclical: In 1980, prices ranged as high as a dollar twenty cents a pound, a big jump from depressed prices a few years earlier of fifty cents a pound.

It is the most versatile of all metals, with literally thousands of uses. It is widely employed in the communications, construction, and transportation industries, and for numerous household items such as toasters, lamps, pots, and jewelry. Its versatility derives from a unique combination of valuable physical properties; it has better electrical and thermal conductivity than any metal other than silver. About half the world's copper production is used for electrical purposes, since it is very ductile and malleable, and of course, cheaper than silver, the only other metal with better conductivity. Copper is the principal ingredient in solar energy collectors. Copper and its alloys, brass and bronze, are also the most esthetically pleasing commercial metals, the only ones with natural, warm, glowing colors. They are extremely resistant to corrosion and virtually indestructible under normal conditions. Copper's durability is legendary. Copper pipe installed 5,400 years ago to carry water to the pyramid of the Egyptian pharaoh Cheops is still operational today.

have gotten up from my cot that early morning in 1960 if I had known all that lay ahead: three more days of excruciatingly arduous climbing before I actually saw and sampled the Ertsberg for myself; a harrowing seventeen-day trek back to the coast during which our native porters deserted us and we were nearly forced to abandon the expedition; and years of political and economic uncertainties, during which it often seemed as if I would not only be one of the first outsiders to see the Ertsberg but the last. The legal, financial, and commercial agreements to mine the Ertsberg eventually filled three fat volumes of several thousand pages, weighed 32 pounds, and occupied 16 inches of library space.

Even when development of the Ertsberg finally began in the late 1960's, the project was nearly doomed by often seemingly unsolvable construction problems. As Dozy once remarked, "It would be hard to imagine a more awkward place in which to find an ore deposit." A 63-mile access road had to be laboriously carved through the hostile terrain, necessitating bulldozing millions of cubic yards of dirt that weighed nearly as much as the Ertsberg itself. Tunnels had to be driven through the middle of two mountains. Complete logistical support for the roadbuilding had to be provided by a fleet of six helicopters. The largest tramways in the world had to be built to move the ore from the mine to a mill in a canyon 2,500 feet below. And the world's longest slurry pipeline had to be assembled to move the copper concentrate up and down the side of the mountains to a coastal port. The entire support infrastructure had to be imported from the outside world. The trade press aptly called the venture "Freeport's Mission Impossible." One journal termed it "one of the most remarkable engineering challenges faced by the world's mining industry." The company's investment in the project figured in current dollars will soon exceed $500 million.

One of our biggest challenges was to find ways of helping the simple Stone Age people in the area adjust to their sudden confrontation with the technological and social complexities of modern Western civilization. Our goal was to improve their austere living conditions without disturbing the valuable qualities of their traditional existence.

I relished the view of the Carstensz Top as I ate breakfast and prepared for another day of climbing, but within an hour, as so often happens in the New Guinea mountains, the gleaming glacier had disappeared, blotted out by thick layers of advancing clouds.

Chapter One

His friends, Jean Jacques Dozy recalled, thought the idea of forging into the mountainous interior of the world's most trackless wilderness was "sheer folly." The company he worked for refused to be publicly identified with the expedition. But that had not deterred Dozy and his associate A. J. Colijn. The Carstensz Top, the highest peak between the Himalayas and the Andes, had never been climbed and they were determined to be the first.

Dozy was telling me about his discovery of the Ertsberg over a drink in an apartment in The Hague one winter evening in 1959. A slight, gracious individual fluent in several languages, Dozy had a remarkably retentive memory for the details of his trek 23 years earlier to what he called *"Naar de Eeuwige Sneeuw,"* the Land of the Eternal Snow. Finding the Ertsberg, he said, had really been an accident, the product of a few casual geological studies he had made along the way "to lend a degree of scientific dignity to what was the pure sport of mountain climbing."

Though Dozy's and Colijn's employer, Shell Oil, had disclaimed public sponsorship, the company did consent to loan the two mountaineers what turned out to be the key to the expedition's success, a two-engine Sikorsky S-38 aircraft piloted by a Lieutenant Franz Wissel of the Dutch navy. The main reason several earlier efforts to climb the Carstensz Top had failed was lack of air support. Climbing the Himalayas is no Sunday afternoon stroll, but it is at least possible for climbers and their bearers to carry all of their supplies with them. The dense swamps and jungles of Southern New Guinea are so maddeningly resistant to passage that it is nearly impossible to traverse them even without supplies. An expedition simply cannot carry enough food and other provisions to sustain itself. The Sikorsky enabled the Dozy-Colijn party to be replenished from parachute drops at strategic points along the route.

With 38 local porters, Dozy and Colijn set up their first camp on a small sand spit at the mouth of the Ajkwa River on the south coast of New Guinea in late October, 1936. On December 5, after 57 days of fighting through the swamps and jungles, the expedition—including Wissel, who, after completing the supply drops, had made a forced march overland from the coast to join his companions—reached the high point on the north wall of the Carstensz Top and explored the full east-west extent of the glacier.

From the near vertical north wall, they became perhaps the first non-airborne white men to gaze upon Baliem Valley, the great central valley of Western New Guinea. At the time, nothing was known of the Baliem Valley, but in 1938, an American explorer on a botanical and zoological expedition discovered what he called a lost civilization in the valley. One of the very few fertile areas of Western New Guinea, the valley was extensively cultivated with the aid of an elaborate irrigation system. It was inhabited by more than a half-million Stone Age people known as the Danis who were completely isolated from the rest of the world.

Six days before reaching the glacier, Dozy and Colijn had pitched camp at the base of a soaring 2,000-foot headwall. They spent several hours the next day scaling it. "It was the toughest individual climb of our entire trip," Dozy told me.

He continued: "As we emerged through the scrub growth at the crest, we saw the Carstenszweide [a meadow carved by glaciers near the Carstensz Top's upper slopes] stretching out before us. Close by on our right was the Ertsberg. This exposure of ore was an impressive sight. It was a black wall rising up several tens of meters with irregular splotches of green copper minerals and a cap of light-colored marbleized limestone.

"I walked around the base of the steep-walled outcrop and as far as I could determine it was all copper ore. Its character appeared uniform throughout except for certain patches on the walls where variations of weathering had resulted in the formation of secondary copper minerals. I made no attempt to climb the cliff although I recall there were several places where it could have been scaled for several meters without too much difficulty.

"At the base of the Ertsberg there was an accumulation of angular blocks of ore which had fallen from the cliff.

These I attacked with my field hammer, breaking off pieces which revealed bright sulphides (of iron and copper) under a thin skin of surface weathering. My recollection is that sulphide minerals, particularly chalcopyrite, were visible in every piece I broke off; not small disseminated blebs (or specks), but massive sulphide. Each broken piece of ore I picked up was damned heavy, and having previously accumulated 80 or more rock specimens, all of which would ultimately have to be returned to the coast on someone's back, I was not inclined to collect a complete suite of Ertsberg minerals. I am sure that I did not give any serious thought to what the average copper content might be. I must confess that 24 years ago it never occurred to me that this deposit could be of economic importance."

Dozy reported the discovery to his superiors at Shell, but considering the remoteness and inaccessibility of the location and Dozy's own lack of enthusiasm, it is not surprising that they had no interest in pursuing it. Dozy wrote a geological account of his trip, which included a brief reference to the Ertsberg. He noted that assays of ore samples he had brought back showed high-grade occurrences of copper and iron and some traces of gold. The report was published by the University of Leiden in the Netherlands in the summer of 1939. However, it received very limited circulation, for several months later Holland was overrun by the German army.

The few copies that had been printed lay unnoticed in local libraries until 1959, when Jan van Gruisen, managing director of a Dutch mining concern called Oost Borneo Maatschappij N.V. (OBM) or East Borneo Company, be-

came interested in potential nickel deposits in Western
New Guinea and ordered a search for any papers about
the island's mineral deposits. The search turned up Dozy's
forgotten report. Van Gruisen had not accorded it any
particular significance, though he did take the routine
precaution of applying to the Netherlands government for
an exploration concession covering 100 square kilometers
(36 square miles) surrounding the Ertsberg.

While on a business trip through Europe during August,
just a few weeks after Van Gruisen found the report, I
stopped by to see him. I had known him for years and once
I had assisted the East Borneo Company in the exploration of nickel deposits on the Indonesian island of Celebes.
More recently, we had corresponded about another nickel
project OBM was considering. Rather casually, it seemed
to me, he handed me some excerpts from Dozy's paper
and asked for my reaction.

My reaction was immediate. I was so excited I could
feel the hairs rising on the back of my neck. The excerpts
contained not only Dozy's description of the Ertsberg but
a general analysis of the area's geology. The sedimentary
deposits, mainly limestone, that covered much of the
Carstensz range appeared to have been extensively inter-
laced with rich ore deposits, especially copper. (Like all
metals, copper has igneous or volcanic origins. Ironically,
as a petroleum geologist, Dozy had been primarily interested in the region's sedimentary deposits, which are the
source of oil.) The Ertsberg, it seemed likely, might be
merely the surface outcropping of even larger copper
deposits underground.

Freeport, my employer, was principally a producer of sulphur, but Dozy's report seemed ample justification to consider a diversification into copper.

When I told Van Gruisen I was indeed interested, he said that his company, which lacked the funds to undertake the effort by itself, would consider a joint exploration venture with Freeport. After examining Dozy's complete report, including maps and photographs, which reinforced my first reaction, I cabled Freeport's New York office and received authorization to spend $120,000 to mount an overland expedition to sample and evaluate the Ertsberg. During the fall, Van Gruisen and I worked on organizing the logistics and drafting a formal contract, which was eventually signed on February 1, 1960.*

Though necessary, the preparations by this time had become a bothersome chore, for I was finding it very hard to concentrate on such dull matters as equipment lists. My own expedition, I was well aware, could easily turn out to be sheer folly, too, but I also knew I must see the Ertsberg for myself or die trying.

Almost everything about the trip, it later seemed, was anomalous, improbable, and bizarre, but it began during April, 1960, in a very conventional fashion: fittings for safari and alpine clothing at Abercrombie & Fitch in New

---

* East Borneo Company, which was later acquired by another concern, did not participate in the eventual development of the Ertsberg, but a company established by East Borneo retains a 5% interest in Freeport Indonesia, Incorporated, an unconsolidated, 81%-owned subsidiary of Freeport Minerals Company, which is mining the Ertsberg.

York City, and repeated puncturing of our arms and rumps at New York Hospital. The concentrated series of injections and inoculations to which we were subjected included cholera, yellow fever, smallpox, tetanus, typhoid and paratyphoid, typhus and scrub typhus. Protection against some of these dread diseases required two and sometimes three hypodermic applications. Some had side effects, and throughout the month I suffered a constant low-grade fever. Further, having reached the age of 50, there was some doubt as to my ability to scramble up the rugged mountains of New Guinea. The usual test is simulated exertion, jumping 25 times on the right leg and then 25 times on the left leg, followed by check of pulse, blood pressure, and so forth. In my case Dr. Robert Watson used a special three-step testing mechanism. After being completely wired for an electrocardiogram, I trotted up and down the steps 25 times and then lay down while a reading was taken to check my heart. Apparently I passed with flying colors. Two months later, I often thought of that three-step test as I rested my weary body after a quick ascent of 4,000 feet or more through mud, roots, and a tangle of vines.

By the end of April, John Bowenkamp, a Freeport mining engineer and longtime friend whom I had asked to accompany me to New Guinea, and I were completely outfitted with jungle boots, ice creepers, fleece-lined jackets, shorts, and rain suits. The yellow pages of our International Certificates of Vaccination gave proof of our immunity to at least half of the ills to which man is heir.

The bulk of our baggage of jungle and Alpine equipment was crammed into heavy-duty canvas duffel bags and sent by air freight to Biak, a small island north of the main island of New Guinea. The freight forwarder confirmed their arrival on May 12th and three days later we left New York at 0900 on United Airlines flight #100. At noon San Francisco time we checked in with Pan American for the evening flight to Tokyo.

On the afternoon of May 20th, after a two-day stop in Tokyo, John and I boarded a KLM plane to Biak which had left Amsterdam late in the evening of May 18th. As soon as the aircraft was at cruising altitude, we were handed a five-page menu listing the gastronomical delights that awaited us. It was quite evident that the passengers on the North Pole route from Europe to Asia had devoted most of their time to the serious business of eating and drinking: a nightcap and snack over the north of Scotland; breakfast on May 19th before touching down at Anchorage, Alaska; a mid-morning fill-in over the Aleutians; lunch extending into May 20th crossing the dateline; and tea in the afternoon before landing at Tokyo. We and the weary travelers from Holland still had to face cocktails somewhere between Iwo Jima and Taiwan; dinner east of the Philippines; and the inevitable late snack over the Caroline Islands. It was pleasant to eat one's way around the world—and the fare was in rather distinct contrast to that which would be available in the vicinity of the Ertsberg.

The plane landed at Biak at 0130 on May 21st. We somehow had expected to step out into cool night air, but the warm, moist steam bath that greeted us left no doubt that we were in the tropics. Tranquil now, Biak was the scene of bitter fighting during the Second World War. Its beaches are still covered with debris from crashed B-25 bombers and landing craft used during the American invasion in 1944. Though the jungle is reclaiming the battle areas, in wandering around over the next few days I still came upon rusting bayonets and bullet-pierced helmets. In the limestone hills overlooking the air field are the crumbling remains of Japanese bunkers and gun emplacements and several natural caves that served as hideouts for thousands of Japanese soldiers.

At the airport, the traffic agent for Kroonduif Airlines, the local subsidiary of KLM, helped us through an informal customs and advised us that our charter flight to the south coast of New Guinea would take off at 0700 on the 22nd. We asked him where we could pick up the baggage which had been sent out from New York three weeks previously. Up to this point, the way things had been clicking off right on schedule seemed to us clear proof of the care we had taken to plan every minute detail of our expedition. That feeling was instantly dispelled when the traffic agent replied, "What baggage?"

We knew we were in trouble. The small Chinese clothing shops in Biak and even the supply depot at the Dutch marine base were no substitute for Abercrombie's. Finally, after four days of frantic searching, during which we even considered a round trip to Sydney, Australia, we were told our equipment had been located in a freight depot at Idlewild Airport in New York. The two heavy duffel bags were rushed on board the next plane. They arrived on May 28th and our long delay was over.

It was still dark on the morning of Sunday, May 29th, when Captain Loonen of Kroonduif Airlines called for John and me at the Rif Hotel with a Land Rover already half filled with our gear. As the heavily laden car bounced around the corner of a small hangar at the local seaplane base, six Papuan boys were trundling a single-engine De-Havilland Beaver seaplane across the ramp. In a few minutes the plane rolled on wheels down the ramp into the clear blue-green water of Japen Strait and two native boys swam out to disconnect the wheels from the plane's pontoons. At that moment the sun was rising out of the Pacific Ocean. Due east, the first land was the coast of Ecuador, more than 10,800 miles away. In the clear early morning light a group of small islands several miles south of Biak were flat-topped mirages with no apparent connection to the mirrorlike surface of the water.

The Beaver lifted easily into the air at 0658 and flew due south, passing over Japen Island at an elevation of 6,000 feet. From this point the northwest coast of the main island of New Guinea was clearly visible. In the foreground, wet and glistening in the flat rays of the sun, was a vast mangrove swamp extending inland for 15 miles. Beyond the swamp, the central range of the New Guinea highlands rose up to more than 16,000 feet. An hour later the plane crossed the coast at a point called Nabire, climbed to 10,000 feet, and flew southeast following the steep-walled canyon of the Sirimo River.

Soon we were over Wissel Lakes, named in 1936 for their discoverer, the pilot Wissel who was a member of Dozy's expedition. Wissel Lakes, now called Paniai Lakes, is a series of three lakes, the principal one being at an elevation of 5,700 feet and having the form of a rough 10-mile square. It is a beautiful body of water almost completely surrounded by mountains ranging up to 10,500 feet. Adjoining the lake were open fields that appeared to be under cultivation by the Papuans. Through frequent openings in the clouds we could see two small airstrips and several collections of huts along the southeast shore. These were the local headquarters of the Catholic and Protestant missionary groups. Established after World War II, for the most part, these and other missions were just about the only evidence of Western civilization on the island.

The southern watershed of the central highlands beyond the lakes was socked in with a thick cloud cover extending to the Arafura Sea. Shortly before the plane started down through the murk in a flat glide, we had a brief view of the western end of the Nassau Mountains. The tops of several 13,000- to 15,000-foot peaks stood out like islands in a white cotton sea. Sitting in the co-pilot seat, I watched the altimeter drop from 10,500 to 1,500 feet and wondered how high the trees grew in the upland jungles. At an elevation of 1,000 feet the pilot started to swing in a wide circle to the right and away from high ground. He was looking for a hole in the clouds. Finding one he spiraled down in a steep bank and leveled out at 300 feet, which seemed about tree-top level. Big, cumbersome hornbills and beautiful sulphur-crested cockatoos emerged from the forest and turned sharply in terrified flight from our wing tips. Their protesting screeches must have been audible for miles.

The cloud cover dead ahead partially obscured the tops

of high trees and the pilot started another wide circle to the right. Actually, he was now circling not to avoid low-lying clouds, but to determine his precise location by checking the configuration of rivers and streams against a map strapped on his right thigh. After nearly 20 minutes of what appeared to be aimless wandering and tree-top dodging, he spotted a large river, the southern-flowing Minajerwi. Dropping down below the level of the tree branches, he flew along the river, which was wide enough to accommodate our plane should a landing become necessary. Three large dugout canoes suddenly appeared downstream and four paddlers in each canoe looked up in amazement as our plane swept over. We passed a native village on the right bank and could see people running under the palm trees, some toward the river to have a better look at us, and others into the jungle in apparent fright.

The pilot turned and crossed several tidal inlets in the coastal mangrove swamp. Under the left wing we could see in one inlet a small diesel launch moored to a mangrove tree. In the stern of the craft was a bearded white man named Jan Ruygrok, a member of a team we had sent over to do advance reconnaissance. Waving to us from the boat were six Papuans. At 0930 the plane slanted down through a light rainfall and made a smooth landing on a broad inlet referred to on the map as Inaboeka, into which the Mawati River feeds.

Bowenkamp and I transferred to a diesel launch, watched the Beaver take off, and headed inland. The first few miles were along a tidewater basin that cut through the mangrove swamps. Water is the only means of passage through these swamps, which are usually ten miles or more wide along the southern coast. A tree that grows in tidal areas in mud up to one's waist, the mangrove is often 50 feet tall and has a tremendous complex of roots, usually one to three inches in diameter, that grow out of the base of the main trunk in a series of hoops. This effectively creates a high fence around each tree. In a dense thicket of mangroves, the roots make up such a solid mass that walking through, especially when one is already waist-deep in mud, is virtually impossible. It is also dangerous, as the area is infested with crocodiles known to reach lengths of 25 feet.

At the end of the tidewater, close to the northern edge of the mangrove swamp, the launch breasted into the strong flow of the Mawati River. At first, our small craft almost failed to make it through the rapids, but once in the main course of the river, it moved forward easily. The river was now bordered by large palm trees and in open, particularly cleared areas we could distinguish breadfruit trees, bananas, and small clumps of sugar cane.

As the launch rounded a sharp bend in the river, we suddenly saw standing before us on the crest of a 20-foot bank at least a hundred Irianese clad mainly in bright-red loincloths. Behind them was a long row of grass-thatched huts. This was the village of Omowka, where our expedition had established its first camp and from which it would begin the trek north to the Ertsberg.

In the 16 days since we left New York, we had not only traveled halfway around the world but, it seemed, we had gone back thousands of years in time.

Chapter Two

The village of Omowka had about 150 residents and consisted of 60 small rectangular huts with grass roofs. Neatly arranged in three rows running parallel to the river, the huts were raised on stilts about three or four feet off the ground. Also in the village were an open-air building used as a mission school and a large barnlike structure which served as a Catholic Church. An empty Japanese shell case served as a bell to call the faithful. From the size of the coconut palms growing on the bank of the river, I estimated that the village had been inhabited for 10 to 15 years. Omowka did not appear on any map, but a settlement referred to as Watimapare was shown two miles upstream. Apparently it had been abandoned during World War II, when the Japanese occupied many coastal areas of New Guinea. The Japanese treated the Papuans quite badly, I was told, and not long after the Japanese arrival, Watimapare's residents had fled into the jungle. When the Japanese withdrew, the residents decided that rather than return to their old village they would build Omowka.

The Irianese of the south coast seemed tall compared to the inhabitants of Biak. They were dark-skinned, but brown rather than black; they had short, kinky hair, and their teeth were filed to sharp points. At the time, an abbreviated loincloth was the only article of apparel worn by both men and women.

Omowka appeared to be the center of the region's dugout-canoe industry. Canoes were made from trees at least 30 inches in diameter, and the finished product varied in length from 20 to 50 feet. The thickness of the wood left on the sides was less than 1 inch; possibly 2 inches remained on the bottom. A flat area about 12 by 24 inches, with grooves to accommodate bare feet, was formed into an overhang at both ends of the canoe. Either on or immediately under these flat areas there was usually a carving of a man's head, which we were told portrayed an ancestor. The men worked rapidly with their small adzes, which only recently had replaced traditional stone axes. It was quite apparent that unusual skill was required to keep from cutting through the sides or bottom of the canoe as it neared completion. The dugout-canoe production line included three canoes in process, six recently completed and drying on the river bank, and ten submerged in the river for seasoning. Moored to the river bank were at least forty craft that appeared to be in daily use.

Besides canoe manufacture, the daily routine of the men in Omowka seemed limited to hunting in the upland jungle and fishing in the bayous of the tidewater swamp. The women, who seemed to do most of the work, disappeared upriver early in the morning to work small cultivated areas of sugar cane and bananas, to gather edible starch from the sago palm, and to recover large pieces of worm-infested wood. Returning in midafternoon with sugar cane, bananas, sago starch, and chunks of rotting wood honeycombed with perforations, they spent several hours at the community canoe mooring gossiping and removing fat grublike worms from the wood. From observations made at the top of the river bank, I judged that about every fifth worm was popped into a female mouth. The balance went into a container and presumably were served later as hors d'oeuvre to the men of the house.

Our own camp was a building of 20 feet by 50 feet. The floor, made from round poles and flattened sections of bamboo, was three feet off the ground. The protection from crawling things afforded by the raised floor was some-what offset by the large population of community hogs which met thereunder every night in squealing conclave. The structure had been hastily erected by a group of 12 villagers who subsequently shared as compensation a 100-pound bag of rice, which was regarded as a delicacy. This payment was agreed to after several long powwows with the native head man and numerous conversations on our jungle radio with Dr. Feldman, the resident administrator for the Dutch government, whose headquarters were in the village of Kokonao, 50 miles to the west.

The route that began at Omowka was selected only after extensive study of different approaches to the Ertsberg. The easiest access, at first glance, might seem to be by parachute drop. This strategy was quickly rejected, though. Because of the mountainous terrain, an airplane could not have approached closer than 2,000 feet from the ground, subjecting the parachutist to dangerous uncertainty as to what he would be landing on. If he was so unfortunate as to find himself snagged halfway down a sheer incline several thousand feet high, he might never get out, for there is no place within many miles of the Ertsberg where a plane can land.

The choice was reduced to the best land route. Dozy had shown that it was at least possible to reach the Ertsberg on foot. We felt, however, that there might be other land routes that would be quicker and more convenient than his. (See Appendix Note 1 for an account of earlier expeditions to the New Guinea highlands.)

The task of researching the problem was assigned to our advance team. The team's leader was Augustinius Wintraecken, universally known as Gus, an intense, 38-year-old Dutch engineer with extensive experience in Indonesia, especially in handling local labor in remote locations. He not only spoke good Indonesian as well as English, but was quite adept in picking up local dialects. He was assisted by Jan Ruygrok, a former Dutch marine and an accomplished botanist in his early twenties, and by Delos Flint, my top field man at Freeport and a close personal friend. Besides route selection, the advance team's main functions were to assemble and arrange for the delivery of supplies and equipment and to serve as trail breakers.

In February, 1960, Gus had met in Biak with missionaries from the Wissel Lakes area who had maps indicating a series of native trails extending from the lakes a considerable distance to the east. One trail seemed to reach a high and barren limestone ridge which was believed to continue on to the Carstensz Top. After studies of aerial photographs taken of this area by the United States Air Force in 1944, Van Gruisen and I became convinced that an approach from Wissel Lakes running nearly due east for about 60 miles might be an easier means of access to the Ertsberg than the long pull north through the jungle from the south coast.

This strategy found support during a meeting in New York that I had with Victor de Bruyn. While living among primitive tribes in interior New Guinea during World War II, de Bruyn had gained fame sending out radio messages to the American Army on Japanese troop movements. He probably knew more about the central highlands than any outsider. He had not been above an elevation of 8,500 feet in the area east of Wissel Lakes, but he thought the eastward route was clearly preferable to an approach from the south, which he characterized as the most formidable terrain in New Guinea.

Encouraged, we decided to make an aerial reconnaissance. In March, Van Gruisen came out from Holland to join Gus, Flint, and Ruygrok in Biak. A Grumman Mallard aircraft was chartered and the chief pilot of Kroonduif Airlines was engaged to fly it. From Wissel Lakes, the plane headed up the valley of a river discharging into the lakes whose headwaters were in a deep canyon north of Mount Idenburg and a few miles west of the Carstensz Top. It followed the barren limestone ridge which looked so inviting in the U.S. Air Force photographs, turned slightly south of the glaciers, crossed the Carstenszweide and flew over the Ertsberg, plainly visible and casting a long shadow in the early morning light. Continuing on a southerly course parallel to the Dozy route, the plane finally landed near the mouth of the Mimika River at the village of Kokonao.

The reconnaissance flight had been carried out on one of those rare days when the entire Nassau range was free of clouds. The view of the ice fields, the Carstensz Top, and the deep boxlike canyons had been spectacular. But everyone on the trip agreed that the proposed approach to the Ertsberg from the west was not possible.

The Air Force photographs had been taken from an altitude of 30,000 feet and at an oblique angle some distance to the north of the proposed ridge route. What appeared in those photographs to be long, fairly smooth rock slopes were clearly seen from the reconnaissance plane to be extremely rugged mountains, frequently cut by vertical canyons 500 to 1,000 feet deep. Furthermore, the approach to this ridge was guarded by an upland swamp at least 10 miles wide. A route up from the southern coast, everyone decided, was our only choice.

After the plane landed, Van Gruisen discussed possible south-coast routes with Dr. Feldman and Father J. P. Koot, longtime head of the local Catholic mission. They knew of a rough trail parallel to and 30 miles to the east of Dozy's route that began in the village of Omowka and proceeded north along the Mawati River. The main advantage of this trail was that unlike the Dozy track it passed through several villages where local bearers could likely be obtained.

While Van Gruisen was in Kokonao, Father Koot introduced him to a remarkable local Irianese by the name of Moses Kelangin. Moses, Father Koot said, might be of assistance to us in organizing our expedition. Van Gruisen could not have known it at the time, but without Moses we probably never would have reached the Ertsberg.

Moses was a member of the Amungme tribe, a part of the Damal tribe, that lived in the interior highlands along the southern slopes of the Nassau Mountains. The Amungmes, who number about 2,500 and are believed to be of Negroid origins, are quite distinct from the Mimika and Sempa tribes of about 6,500 people that live along the southern coast. The two groups speak very different languages and historically have had almost no interaction.

Because of the terrain, the mountain people have powerful leg and thigh muscles, but are very short, averaging less than five feet; Moses was probably only four and a half feet tall. The reason seems to be not so much pygmoid genetics as poor diet. Constant, often heavy rainfall on the south side of the central highlands so thoroughly leaches out the topsoil's nutrients that cultivation of productive crops is nearly impossible. Small plots of a kind of sweet potato serve as the mountain people's main food source. They eat both the root and leaves, the latter resembling spinach. The highlands are so barren of animal life that their protein intake consists mainly of snakes, worms, and a few small birds shot with arrows. They will eat anything that walks, creeps, or crawls.

The coastal tribes in the vicinity of Omowka village are not much better off. Much of the southern coastal plain is either mangrove swamp or so poorly drained that it is constantly invaded by tidal flooding, which contaminates fresh water wells and kills most edible crops. But fish are reasonably plentiful in the tidal estuaries and limited cultivation of sugar cane and bananas as we saw near Omowka is possible. The staple of the coastal people is sago from the sago palm which grows wild throughout the coastal flatlands. Because their diet is better, the coastal tribes are relatively tall. They have powerful shoulders and arms from paddling and poling dugout canoes, their chief mode of transportation. Yet their legs are spindly and weak; they have little overland endurance. The coastal Irianese we used as bearers in the mountains lasted only a day and a half before returning exhausted to the coast.

The coastal tribes have had at least limited contact with the outside world for many decades. For much of their early colonial rule, which dates back to the seventeenth century, the Dutch concentrated on the more developed and commercially attractive islands of Java and the Moluccas which were a source of spices. Yet following Japanese occupation of many coastal sections of New Guinea, the Dutch substantially increased their presence. Though mainly concerned with the offshore islands of Biak and Sorong and with Hollandia (now Jayapura), a city on the north coast of New Guinea which had been the site of a large Japanese and later Allied base, the Dutch stationed administrators in several large south-coast villages, built schools, and organized agricultural centers. Early Prot-

estant and Catholic missionaries were deterred from seeking converts in New Guinea by reports of cannibalism,
but in the early part of this century, a few opened missions
on the south coast and began introducing Western culture.

In contrast, the primitive mountain tribes until very
recently remained almost totally isolated. Unaware of such
advances as the wheel and iron tools, many still live much
the way early *homo sapiens* did a million years ago. Other
than a few hardy mountain climbers, virtually no outsiders
ventured through the jungles and rain forests into the
rugged highlands. And the mountain people were afraid to
travel to the lowlands, for when they did they invariably
contracted malaria. The coastal tribes had built up some
tolerance to the disease but afflicted mountain people
usually died. Even after mining of the Ertsberg began,
coastal people generally refused to work at the mine site,
while the mountain people generally refused to work at
our facilities on the coast.

Moses Kelangin was very different from his fellow
mountain tribesmen, an individual with far above average
intelligence and atypical curiosity about the world beyond
his village. In 1952, when he was about 30, he made his
way down to the coast and joined Father Koot's mission
in Kokonao where, perhaps because of his long journey,
he was given the name Moses. (The name was especially
appropriate for the role he played in our expedition.) After
eight years at the mission, he learned to speak Dutch
and Malay (the root of the present-day Indonesian national
language) as well as the language of the coastal tribes.
He began giving religious instruction to the coastal people

and carried the faith back to his home village in the mountains, where he came to be revered as a great leader. Later, when I went through the village with Moses, he stopped and delivered a long sermon in a church with a grass-thatched roof and wooden logs as benches.

Though he became somewhat westernized, Moses adhered to his tribe's customs. One of the first things I noticed about him was that joints of three of his fingers had recently been cut off. I found out later that in some villages it was customary, when someone died, to cut a joint off one of the fingers of members of that individual's immediate family. Some older women in his village had practically no fingers left.

Moses was probably the only person in Western New Guinea who could talk to us, to the coastal people, and to the mountain tribes. He was also an excellent guide, recruiter, and labor negotiator. Intense and businesslike on the trail, he seemed very involved in what we were doing. On several occasions when it seemed our expedition was about to fall apart from a lack of bearers, Moses always managed to assemble enough people to get us through.

Father Koot agreed to loan Moses to us for six months. In return, we paid his mission $300 and agreed to let Father Koot have all of our surplus supplies, with which he intended to stock other missions in the area. When I returned to Biak, I bought and sent Moses an expensive watch. I later got word that he was very pleased with it, but still the gift was woefully insufficient. His value to us was incalculable.

The route having been chosen and Moses enlisted, Van

Gruisen returned to Holland in March. Gus, Del Flint, and Ruygrok flew to Sorong, a large town on the extreme western tip of New Guinea that had been selected as the recruitment and supply assembly center.

Supplies and equipment from Holland were already arriving. One of the most important items was a large package containing 50 surplus U.S. Air Force parachutes, camouflaged with green and yellow splotches for jungle operations during World War II, which were to be used for airdrops. Other essential equipment included a large supply of plastic-coated nylon tarpaulins, to be used as tents and supply depots along the trail, and heavy woolen clothing for the porters who would assist us in the vicinity of the Ertsberg. We were fortunate in being able to obtain from the local oil company in Sorong three portable radio sending-receiving sets powered by a bicycle-operated generator. Gus, Del, and Ruygrok packed the supplies and foodstuffs in specially marked containers and prepared lists of items to be flown to Biak's World War II airstrip, the longest in the Pacific, where most of our airdrop resupplying missions would originate.

By late April, 1960, the 33-man advance team was assembled. Beside the three leaders, it consisted of a Dutch-Asian radio operator, two camp foremen from the Celebes and Japen Islands, a first-aid man from Sorong, and 19 Irianese from the village of Sansapor near Sorong, who were to serve as porters in the highlands. On April 22nd, the team and the supplies not sent to Biak for imminent use were loaded aboard the *Seremoek*, a 200-ton coastal vessel. From Sorong, the ship steamed southeast to Kokonao, where the radio operator was put ashore. He would later serve as the expedition's communications link with the outside world. Vincent Croin, a Dutch police officer, and four Irianese trained as policemen were brought aboard. At the recommendation of Dutch officials in Kokonao, they had been engaged to guard the expedition's supply depots.

The next afternoon, the *Seremoek* dropped anchor six miles inland from the coast in the Koperapoka tidal inlet, just west of the Mawati River. For the following two days, a small diesel launch transferred personnel and equipment to Omowka, 15 miles upriver. By the time I arrived in Omowka, the members of the advance party led by Gus and Del Flint (Jan Ruygrok stayed behind to accompany the main body of the expedition) had canoed up the Mawati, established our base camp about two miles north of Dozy's base camp near the village of Wa, and were hacking a trail through the mountains toward the Ertsberg.

On the afternoon of our arrival in Omowka, we were kept busy filling five-gallon tins with rice and other food that would have to sustain us on the trail during the ten days we expected it would take us to reach the first supply depot (one of several along our route) which had been established by the advance party on the outskirts of a small mountain village called Jonkogomo. The supply base had been given the code name Tsinga after the nearby Tsing River. Moses, who after guiding the advance party had returned to join us, was completing the recruitment of our bearer party.

The radio equipment was carefully inspected, as our survival in isolated areas might depend upon its proper operation. The sending and receiving set had a range of about 80 miles and was a compact unit neatly packaged in a canvas carrying cover and weighing 30 pounds. The bicycle-driven generator which served as the power plant for the system weighed 44 pounds and was the heaviest individual piece carried on the expedition. The local boys who carried the generator over the trail to Tsinga and from Tsinga to the Ertsberg were paid with two ax blades, one more than the payment for normal loads.

Items such as ax blades, which we called "inducement goods," were our chief form of compensation to the packers, who had no knowledge of or use for money. Barter economics is much more inconvenient than money economics, and about 10% of our packers had to be allocated to the task of moving our bulky and heavy payroll along the trail.

During our first night in camp, there were numerous tropical downpours. The roar of the rain on our thatched roof woke me several times. I was amazed what a noisy place the jungle could be. From dusk to dawn, there was a constant cacophony of high- and low-pitched croaks from frogs, tree toads, insects, and reptiles. Yet this was the only time on the entire trip I heard any night sounds in the jungle. In the mountains of New Guinea, which are largely barren of animal life, even insects, the jungle is very quiet. On most nights during our trip, however, I was so tired that I could have slept through anything.

Chapter Three

Monday, May 30: The Omowka community landing was a scene of hectic activity. Dugout canoes were gathered in the early morning light around the rear of the diesel launch, which had been loaded with supplies the night before. Irianese women formed a chain to pass baggage and supplies to the men in the canoes. Small tan-colored hounds ran up and down the river edge and a large group of chattering naked children watched intently from the top of the bank. (The hunting dogs of the coastal Irianese are small descendants of the Australian dingo, which 25,000 to 50,000 years ago crossed what was then a land bridge between Australia and New Guinea with their aboriginal masters.)

In addition to John Bowenkamp, Jan Ruygrok and myself, our party included the following: Moses Kelangin, our guide, translator, labor negotiator and certainly the most important man in our group; Molle, our radio operator, an expatriate Indonesian from the island of Celebes who spoke Indonesian, Dutch, several local dialects, and a few words of English; and Takim, an Indonesian of Chinese ancestry from Java who served as our cook.

At 0810 John and I settled comfortably into the bottom of a 50-foot dugout canoe which already contained a half ton of baggage, seven paddlers, a female hunting dog with a litter of puppies, and the five-year-old son of the head paddler and owner of our river craft. The child, dog, and puppies immediately disappeared under a form of low tent made from two or three grass sleeping mats. The lead paddler, who stood on the small flat space on the bow of the canoe, stamped his left heel, the seven paddlers dipped their long oars in unison, and we moved up the Mawati River at the head of a 14-canoe fleet with 44 paddlers.

The skill and endurance of the paddlers were remarkable. Except for occasional pauses to relight the splayed end of a horn-shaped wad of tobacco, which passed as a cigar, they kept up a continuous beat, which at times accelerated to a surprisingly high rate. The canoeing procedure traveling upstream was to keep close to that bank of the river most likely to have a helpful back eddy. To achieve this objective the canoes made frequent crossings from one bank to the other, the time of crossing depending on the judgment of the lead paddler. I noticed no uniformity in the pattern of river crossing by the canoes; sometimes when our group was paddling furiously to cross from left to right another canoe would cut across our stern headed in the opposite direction. During the crossings the lead paddler in each canoe would gradually increase his beat as his craft breasted into the faster current in the center of the river. The beat would decelerate as the canoe encountered a favorable eddy and glided upstream brushing against tall reeds and trailing vines at the river edge.

The lead paddler in our canoe, who wore a Japanese army cap, perhaps to signify his status, was in charge, having been accorded by his comrades the responsibility of making decisions on such matters as when and where we would stop for the night. But he seemed quite content to permit other canoes, at least on a short-term basis, to lead the parade. Relative positions changed continually as canoes passed one another with short bursts of speed. For the paddlers, this was apparently a means of breaking the monotony. For the passengers the trip was anything but monotonous.

The Mawati River in its lower reaches, like most of the numerous rain-fed rivers emptying into the Arafura Sea, follows an extremely meandering course. In the first two miles above Omowka, where the river was about 150 feet wide, the dense jungle on both sides of the river was broken by occasional small clearings where sugar cane and bananas were cultivated. The abandoned settlement of Watimapare was identified by a grove of extremely tall coconut palms gradually being choked out by other growth. Above this point the river narrowed slightly and the canoes moved through a canyon whose light green walls were a tangle of immense trees and hanging vines.

In the early morning and again late in the afternoon there was unusual bird activity in the trees adjacent to the river, particularly from sulphur-crested cockatoos and giant hornbills. The hornbills, which travel in flocks of a dozen or more, appear clumsy in the air. The beat of their large wings sounds like a group of bellows and can be heard for hundreds of yards. The cockatoos are without doubt the noisiest creatures in the jungle and emit piercing screeches as they swoop from one tall tree to another. Numerous smaller birds of many colors were visible and on the afternoon of the second day we saw crossing the river a medium-sized brown bird with yellow breast and long trailing yellow and brown feathers, one of the more common varieties of the bird of paradise.

After four hours of paddling, a stop was made for lunch on a sand bar. The paddlers started numerous small fires of driftwood and threw lumps of sago starch about the size of tennis balls into the coals. When these unappetizing items were burnt to a black char, they were fished out of the fire with a stick, broken open, and eaten like biscuits.

Two hours after leaving the sand bar, the lead canoe pulled into the left bank where there was a row of low lean-tos in a state of disrepair, probably a former hunting camp. The other canoes fell in behind and we were told this would be our first overnight camp.

We took off our clothes for a swim in the swift river while the tents were set up. A curious audience began observing us from the banks. I learned later they were fascinated by the contrast in skin color between our tanned legs and backs and our white mid-sections. During later swims, Irianese would sometimes come up and pinch the white and tanned sections in an effort to understand the disparity. Once, while we were walking in a heavy rain, several mountain tribesmen we met on the trail pulled at the wet khaki shorts that were clinging to our bodies. Having never seen cloth before, they thought the khaki was some curious form of loose skin.

The Irianese quickly set up lean-tos, spread out grass sleeping mats, and started numerous fires for cooking sago balls. Camp construction was a ritual to which we soon became accustomed. First, two stout poles about 10 feet long with Y-shaped tops were driven into the ground some 23 feet apart, and an equally stout ridgepole was laid horizontally between the Y's. To form platforms for sleeping and storing of cargo under what would become a 20 x 20-foot tent, 12 heavy sticks about 2 feet long, also with Y-shaped tops, were driven into the ground in rows parallel to the ridgepole and so arranged that a three-foot corridor would remain in the center. Round poles were run between the Y's of the short stakes, and then small branches, an inch or more in diameter and 7 feet long, were laid crosswise between the poles, thereby forming two platforms about 7 feet wide and 20 feet long separated by the corridor. The small branches of one platform were covered with a three-inch layer of ferns and leaves on which six bedrolls would be spread. The other platform accommodated the baggage and supplies. When the details of the ground floor were completed, a 20 x 22-foot plastic-coated nylon tarpaulin was thrown over the ridgepole and its edges were then tied to the ground so as to leave an air space at the level of the sleeping platform.

Camp 1 was located in a typical rain forest of large trees, some of them six feet and more in diameter and towering up for at least one hundred feet to the first set of branches.

The forest is roofed over with what appears to be a solid mat of leaves. Because the sunlight cannot penetrate, the only underbrush is small leafy plants and ferns. In contrast to the steaming heat of the open lowlands, where the temperature is a fairly constant 85°F. to 90°F. with very high humidity, the rain forest is cool and comfortable.

At dusk, after the radio equipment had been set up, we went on the air. Because of intervening high mountains, we were not able to talk to the advance party, but we did raise our operator in Kokonao after a few minutes of calling and learned the good news that on the previous day the advance party had reached the Carstenszweide, the meadow near the Ertsberg. We also talked with Kokonao about the timing of the next airdrop at Tsinga, several days travel to the north.

Substantial supplies had already been air-dropped at our principal supply points en route: Tsinga, Base Camp, and the Carstenszweide. In all, 36,000 pounds of cargo were delivered by air. Parachutes were used for the Carstenszweide drops because of the tall mountains nearby, but most of the deliveries to Tsinga and Base Camp were "free drops" without parachute. The plane, a twin-engine Pioneer, would approach the target area, a cleared field denoted by a smoky smudge fire, in a steep downward slide. At an elevation of only 100 feet, the pilot would apply full throttle and pull the plane into a steep climb as the crew shoved burlap sacks through the open door. Three or four passes—and strong nerves and steady hands for the pilot and crew—were required to drop one plane load.

Breakage was surprisingly low—less than 2%—for everything had been heavily padded and wrapped and re-
wrapped and then the parcels packed in sawdust. One
bundle dropped at Base Camp did crash into the trunk of
a large tree, scattering cans of condensed milk over a wide
area, and during a parachute drop on the Carstenszweide
a 50-gallon steel drum broke loose from its parachute and
plummeted like a bomb into a heavy swamp. The drum
split open and gouged out a four-foot crater that immediately filled with water. When it was dragged out of
the hole, it was found to contain the expedition's entire
supply of sugar. For the next month, our coffee and tea
were sweetened with a thick, slightly dark mixture of sugar
and bog water.

In planning the air drops, we were worried less about
breakage than about a religious phenomenon common
throughout the South Pacific known as "cargo cultism."
Cargo cults vary widely, but most consist of usually pagan
rituals which cultists hope will result in an abundant flow
of cargo or material goods. The cults first sprang up in New
Guinea when Westerners arrived in huge ships carrying
exotic material. The local villagers, whose horizon of material goods was limited to stone axes, dugout canoes, grass
huts, and whatever food had been gathered for the evening meal, were amazed and envious. Their amazement
and envy were intensified during World War II when ships
and airplanes unloaded thousands of tons of weaponry and
supplies at coastal bases.

The villagers had no sense of the processes of scientific
experimentation, analytical reasoning, and industrial man-

ufacturing that lay behind this cargo. Magic, miracles, and myths are strongly embedded in Melanesian culture, especially in such primitive areas as New Guinea, and the villagers assumed that the cargo must be of supernatural or divine origin. The outsiders had been given the cargo by a deity through some secret ritual. Their relatively deprived local culture, the villagers reasoned, was due to the fact that they lacked the Westerners' secret. Over the years, local leaders developed many rituals of their own which they hoped would bring forth material affluence. Sometimes, when tribes became convinced that the millennium was close at hand, they burned their huts and crops and sat on the ground to wait. In a few instances, the population of entire villages starved to death.

Our air drops, fortunately, had no such tragic consequences, but cargo cultism remains embedded in Irian culture even today. Tribesmen living in the vicinity of the Ertsberg have had difficulty comprehending the modern mining enterprise that has materialized in the area over the past few years. There has been some friction between Freeport and tribesmen who feel that the company could only have obtained such incredible treasures as helicopters through a cargo secret which it is unreasonably withholding from them.

Shortly after we finished talking to Kokonao, a heavy rain began to fall. The roof of the rain forest, it was apparent, had many leaks and we were soon drenched. Near midnight one of the tarpaulins, put up with a slight sag, developed a water-filled depression of bathtub size, broke its connection with the ground and spilled a small waterfall on two members of the party, forcing them to squish about in search of a more comfortable bed.

Tuesday, May 31: The camp was stirring before dawn. After a breakfast of hot tea and rice prepared by Takim, the tents were struck and at 0735 the fleet moved out into the current. Within a short distance from Camp 1 the river narrowed and its character changed from a swift, even flow to a series of rapids between deep pools. The oarsmen became polers in the rapids and on several occasions leaped into the shallow water to become draggers. We encountered several partial log jams and progress was halted from time to time as a way was cut through a tangle of branches.

During the day we spotted several new varieties of birds. On one sandbank we noticed the three-toed tracks of a cassowary. The cassowary is a large, flightless bird which can disembowel a man with its hooklike talons. The Irianese use its bushy tail for headdresses and other ornaments and its two-inch talons as tips for wooden spears. From time to time a paddler would leave his oar in his canoe and head off into the jungle with a spear and a small hunting dog. The others, meanwhile, were doing less and less paddling and poling as forward movement now consisted principally of dragging canoes through a series of shallow rapids, an exercise in which all of us participated.

Early in the afternoon the lead canoe pulled into the bank and the others tied up astern to unload cargo. We could have proceeded further but we had reached Camp 2 established by the advance party a month earlier. We were able to utilize their tent poles and sleeping platforms, which were put in order with a new cover of leaves and ferns.

Later that afternoon, the missing canoemen, turned hunters, emerged from the jungle in groups of two and three and laid their game on gravel by the side of the river, where it was skinned and hacked into pieces for cooking. The addition to our food supply was considerable. The bag included a ten-pound river fish, two wild pigs, and three small tree kangaroos. That night we all had meat but the hunk of tough gristle I found in my rice was not much of a complement to the accompanying sardines and corned beef. For the Irianese it must have been a vast improvement on those burned cakes of starch. They showed their enjoyment by singing songs long after dark.

Wednesday, June 1: There was no rain during the night but a lot of wind. By morning, the campground was littered with large leaves and palm fronds. As we moved up the river there was a gradual change in the terrain. The jungle became a more open woodland, the trees less tall, and for the first time the low foothills could be seen in the distance.

During the last hour of the previous day's trip it had become quite clear that our canoes were nearing the end of their use as a conveyance for human cargo. This morning it was easier to travel on foot on the broad areas of gravel on both sides of the river than to attempt to move in and out of canoes being hauled up rapids. As a conveyor of cargo, however, the canoe remained a most efficient medium even when being dragged over a shallow, rocky bottom. The cooperative effort of two or three rivermen could still move a thousand pounds. When the transfer was later made from canoes to the backs of the mountain tribesmen, the necessary number of human conveyors for a thousand pounds of baggage increased from three to at least thirty.

After two hours of walking along the gravel banks, we came to a point the Irianese called Praubivak, literally "canoe camp," or the end of canoe travel. In our case, however, it was a transfer point. The loads were moved from the 40- and 50-foot canoes to 20-foot craft, which were dragged for another two miles upriver to a final impassable series of rocky rapids. We unloaded the canoes and set out on foot.

At Praubivak a trail turned right into the jungle and away from a broad left-hand sweep of the river. About a dozen of us entered this shortcut in Indian fashion, single file. The path was over sandy soil protected from the morning sun by dense forest cover. The going was easy, but I had not progressed 20 yards from the river bank when my feet became entangled in a creeping vine and I fell heavily on my face on the jungle floor. Fortunately, those who preceded me were unaware of my embarrassing plight and the Irianese who followed me were sufficiently polite to avert their eyes. Though I later climbed through some of the most adverse terrain in the world, I did not suffer another fall until I slipped coming down a steep, rocky gorge some six weeks later as we were making our way back to civilization.

In the next hour we saw several huge trees and many unusual jungle plants, but no orchids, which I had been told would be plentiful. The three-toed tracks of cassowary birds were much in evidence and on one occasion we had a brief but magnificent view of a small group of crown doves, known as *kroonduif*, the largest member of the dove family. They are gray in color and about the size of a small turkey, and their heads feature a prominent crest of feathers arranged like a spreading peacock tail.

Shortly before noon we emerged from the jungle into an area of cultivated fields in which taro (a large tropical plant with edible roots), sugar cane, sweet potatoes, and bananas were growing in competition with native grass which often exceeds six feet in height.

We were now in Belakmakema, where we first met the pygmoid mountain tribesmen. That the mountain people had had almost no contact with the coastal people was quite apparent, for the coastal people who had accompanied us to this point showed as much interest and curiosity in them as we did.

In my many encounters with the mountain Irianese over the next few weeks, some in open clearings such as Belakmakema, where we were under observation as we approached, and others in heavily forested areas, where the meetings were often sudden, I was invariably impressed with their friendly nature and open generosity. They always expressed amazement, the women by uttering little high-pitched cries, and the men by responding with a loud laugh. I was never sure whether the laughter was occasioned by my height, 6 foot 3 inches, or my usual disheveled appearance.

Upon meeting, the first movement by the mountain

Irianese was to extend the right hand. The women would only touch fingertips to fingertips, but the men used a form of greeting which can best be described as knuckle-snapping. The man who considered himself the older of the two grabbed the second knuckle of the index finger on the right hand of the younger man with the second knuckles of the two middle fingers of his right hand. After getting a firm grip, the hand was suddenly pulled back, making a loud snapping noise when the knuckles of the two middle fingers of the older man came together. This maneuver was repeated three or four times while each of the participants maintained a wide grin. The next item of etiquette was an offering of a sweet potato or a banana which was extracted from a woven net bag suspended from the brow and slung over the back. I learned to respond by offering a small Dutch chocolate bar from the supply I carried in one pocket of my khaki shirt.

The men wore no clothing other than a curved bright-yellow gourd that covered the penis. Known as a *koteka,* it was from 12 to 20 inches long, about one and a half inches in diameter at its base, and tapered to a point at the top. It was held in a near vertical position by two strings tied to the lower end of the gourd and then around the waist. In recent years, the Indonesian government has been trying to discourage wearing the gourd, which it regards as a rather embarrassingly graphic symbol of backwardness, and to encourage Western clothing. Nevertheless, the gourd is still common in the Western New Guinea highlands.

Most observers have ascribed a phallic symbolism to

the gourd. The Irianese do accord it decorative significance, and the typical man will have a wardrobe of several sizes, small ones for battle and large ones for festive occasions. But it is also very practical. Considering the rough country the mountain people have to traverse, the gourd is probably the most effective available form of protection.

The mountain women are even smaller than the men. As a result of continuous toil, the bearing of children and a steady diet of starch, they have thick, ill-formed bodies with protruding bellies and sagging breasts. They were clad in an abbreviated piece of woven grass about 10 inches long and no more than 4 inches wide. The skirt was tied around the lower buttocks in such a manner that, while I never saw it happen, the garment always appeared about to fall off.

The women had few ornaments other than a small piece of fur from a cuscus, a small opposum-like marsupial common in the highlands, that was worn at the neck, and strings of beads made from nuts and berries. The men, however, had a wide assortment of personal adornment. The septum of the nose, which was always pierced in the male child, accommodated a variety of articles such as shells and even a small tube of bamboo or rattan, but the most popular was a curved boar's tusk which lent the maximum degree of fierceness. Most of the men wore bands of plaited grass on the upper arm above the elbow and at the wrist. Their necklaces included strings of beads, shells, and the teeth of small animals. On festive occasions, the men wore headdresses, either plumes of cassowary tail feathers or a crude hat made of cuscus fur.

The woven net bags, carried by both men and women when traveling on the trail, contained a supply of food and sometimes all their worldly goods. Things I saw extracted from these bags included fire-making equipment, consisting of a split stick held apart by a stone, fibrous bark for tinder, and rattan strips for friction igniters; stone knives and axes, the mountain tribesmen's only mechanical implements; tobacco grown in their gardens; and a rather strange package of woven pandanus leaves which opened up into a 2- by 4-foot envelope split on two sides to serve both as rain gear and a sleeping mat. The women used the net bag to carry small babies or even a piglet and sometimes both at once. Pigs and children are held in approximately equal regard by the mountain natives. Del Flint once observed a mother nursing her child on the left breast and a piglet on the right.

The pig, which was probably introduced to New Guinea by the earliest settlers, has a special place in the culture of the mountain Irianese far beyond its obvious value as a protein source. It is a universally accepted medium of exchange and the most common form of wealth. When we were there, four pigs equaled one wife and two steel ax blades. Reparations to settle tribal disputes are usually made with pigs. Pigs are roasted luau-style on the most important ceremonial occasions. New Guinea cannibals refer to humans as "long pigs," which they cut up and cook in a similar manner. Human flesh is regarded as juicier than relatively dry pig flesh.

Camp at Belakmakema was pitched early in the afternoon on the river bank and adjacent to a deep pool which proved to be a good swimming hole. While I swam, attracting a large audience of Irianese, Jan Ruygrok and Moses counted the available packers. Though 80 were required for the journey north, only 55 men could be found.

The shortage was due principally to our delayed arrival because of the baggage incident at Biak. Several of the mountain people, who previously had been contracted by Moses and were waiting at Belakmakema, had given us up for lost and had returned to their mountain villages. Others had begun using their idle time to start the construction of huts near the garden plots. The Dutch were trying to activate Belakmakema as a settlement for the mountain tribesmen because the area was more amenable to cultivation than villages higher in the hills. We found that once these people had become engaged in a community building project no form of reward was sufficient to induce them to stop construction and accompany us on the trail.

Moses, though, quickly demonstrated his resourcefulness in dealing with labor problems by sending scouts to nearby areas where packers might be recruited. He instructed others to start digging food from the gardens so that supplies would be available if and when we were ready to depart.

Thursday, June 2: The first order of business in the morning was rearranging our baggage to reduce our bearer requirements. At Omowka the baggage had been carefully prepared in 5-gallon tins, each weighing 15 kilos (33 pounds) and coded with a red, yellow or green stripe to identify its contents. By adding more rice and other foodstuffs to each tin, which increased the weight to 16 kilos or a fraction more, we were able to eliminate five loads. After combing the cargo for less essential items, we cut three more loads.

Still further reductions were made possible by the simple device of advance payment. Our baggage included 13 bundles of the inducement goods that were to be used to pay off our bearers, some of whom had already packed for the advance party and had returned to assist us. We had intended to distribute these goods in Tsinga, about 30 miles and a week's travel from Belakmakema. An earlier distribution, it became apparent, could cut our porter requirements.

In an afternoon conclave lasting more than four hours, Moses negotiated the payments. Everyone who had packed for the advance party received one ax blade, one parang (a kind of machete), a small mirror, a little salt, some shag tobacco, and two folders of Italian cigarette papers which were greatly favored over the local product, the leaves of the pandanus palm. (In contrast to Western habit, the Irianese like to bite a hole in the middle of their cigarettes and then light both ends.) Each man who agreed to accompany us to Tsinga was given a prepayment of one more parang. The result was to reduce our loads of trade goods from thirteen to five, thus cutting our porter requirement by eight.

While Moses negotiated, Ruygrok enlisted ten coastal Papuans as additional bearers. The net gain, though, was only five. While the mountain people were able to feed themselves en route, the coastal people, with their different diet, required imports. Four of the ten would have to act as food bearers for their own group. We suspected that the additional capacity might be even less than six, for we were uncertain just how well the coastal people would perform in alien mountain terrain.

Before nightfall every man who had agreed to accompany us had selected his load and had spent considerable time in preparing it for the journey. This included a rather elaborate harness made from split rattan, allowing the packer to carry his 16-kilo load on his back supported by a band passing over his forehead.

In our evening talk with Kokonao I dictated two cables, one to my boss in New York, Bob Hills, then the president of Freeport, reporting on our delayed progress, and the other to my secretary, Adele Roach, asking her to arrange for an appropriate floral present to my wife on the occasion of our twentieth wedding anniversary, two weeks hence.

Chapter Four

Friday, June 3: Up to this point the intrepid foreign travelers had the feeling that they were really roughing it in the wilds of New Guinea. Actually, most of our time had been spent moving with luxurious ease through picturesque country as passengers in a dugout canoe. Our physical exertion had been limited to a few hours of walking over level, sandy trails in the jungle or along the gravelly banks of the Mawati River. Today, however, things were due for a change.

If we had known what actually faced us, I am sure we would have turned around and headed back to the coast. I learned some two weeks later that Del Flint, as a member of the advance group, had been seriously inclined on several occasions to send out a message advising us not to make the attempt to go further because he was convinced that we would be unable to get through. As we soon would, he found that with each day of advance the difficulties multiplied. Having conquered what he was certain was the ultimate, he found that the next day's travel invariably presented a new and greater challenge. Yet he finally had concluded that we would not believe any messages of warning and would insist on finding out for ourselves. In this respect, he was absolutely correct. Having built up such a head of steam over the previous eight months, I was not going to be dissuaded from continuing by anything less than a complete physical collapse.

A light rain was falling in the morning as John and I slung canteens over our shoulders and started up the right bank of the river behind two mountain tribesmen who were loaded with Abercrombie & Fitch knapsacks, and a collection of Takim's pots and pans. We were at least an hour in advance of the main body of the expedition, most of whose members were gainfully occupied well after our departure with the usual chores of breaking camp. As the slowest members, we didn't want to be too far behind by the end of the day.

The trail between Belakmakema and Tsinga was well established and, by New Guinea mountain standards, relatively effortless. To unacclimated outsiders, however, it gradually became grueling. In seeking the best route along the now narrow and shallow river, our guides frequently crossed from one bank to the other. Following, we jumped from stone to stone to keep our feet dry and preserve the feeling of physical comfort that had characterized our journey so far. Our third crossing, though, was at hip depth, and I gave up trying to keep anything dry. Rivers, streams, rain, and mist later would become such a ubiquitous part of our lives that it almost never occurred to me to try to keep my clothes dry on the trail. I felt fortunate if I had something dry to sleep in.

The trail turned from the Mawati valley and headed northeast toward the watershed of the Otokwa River, also known as the Tsing in its upper reaches. At the start the climb was gradual, but soon we encountered a series of steep terraces 50 or 60 feet high. Each left us gasping for breath at its crest.

At these pauses we noted that our two guides were busily engaged in scraping things with knives from their arms and legs. The things, we soon realized, were leeches about the size of inchworms. We thought we were somehow immune until that evening when I took off my boots. My socks were soaked with blood. Though I hadn't felt them, leeches were all over my legs. Bloated from having gorged themselves, they were easy to remove. They practically fell off. But, as I found out the next day, when they first become attached one has to either scrape them off with a knife or force them to let go with a cigarette butt. The slight incision they make in the skin continues to bleed long after they have been removed. Though they were in less abundance than on this first day, we continued to run into leeches for the next few days, even at elevations over 5,000 feet.

As we slowly climbed the sloping foothills, we began to see numerous varieties of orchids, some on trees and others on the ground. The vegetation was less dense now, but occasionally we saw huge trees. One measured 35 feet in circumference near its base and soared over 150 feet into the air.

Four hours of hiking brought us to an abandoned native hut where Molle, Takim, and some of the packers who had passed us earlier were resting and munching on dry crackers. I was exhausted and asked Molle what the afternoon travel would be like. He said that we were over the worst of it and that it would be easy going from here to Camp 4, where we planned to spend the night. His remarks must have been intended to keep my spirits from lagging, for the going that afternoon was far more arduous and tiring than it had been in the morning.

At midafternoon we reached the crest of the ridge between the Mawati and Otokwa rivers at an elevation of 2,500 feet. We found Molle and some of the packers waiting for us in another partially fallen-down native hut in the midst of a small clearing in the forest. While I was sitting in the hut, I noticed a dozen or more large white eggshells, at least the size of turkey eggs, impaled on spines extending out from the trunk of a nearby tree. There was a small hole in both ends of each egg and obviously the contents had long since been consumed by the Papuan who had originally occupied our present shelter. Molle said these were the eggs of a fairly large bird which he called a "jungle chicken." (See Appendix Note 2 for some information I later uncovered about these curious eggs.)

By this time we were bringing up the rear of the expedition and Molle was obviously concerned with our debilitated condition. He said that it would be all downhill from here to camp, which was less than an hour away. I figured that one hour of travel was the limit of trail-slogging my weary muscles could endure.

This time Molle's prediction was accurate. After stumbling downhill through the now-heavy rain for almost exactly an hour, I was delighted to see across a narrow ravine two plastic-coated nylon tents with an adjoining small thatched-roof cookshed. It was Camp 4.

On my arrival, Takim handed me a mug of hot tea laced with condensed milk. I was amazed how quickly this brew revived me. I dug out my personal baggage from the muddy platform under one of the tents, spread the roll, removed my sodden clothes, climbed into a fresh shirt and pair of shorts which had been carefully wrapped in a plastic bag in my knapsack, and promptly went to sleep.

I had been asleep for an hour or so when I was awakened by a noise like a pack of hounds pursuing a fox. The barking sounds were coming from the mountain tribesmen who, directly behind our camp, had located and killed a 14-foot python. It was a magnificent snake and, thinking of the spectacular color picture it would make, I suggested to Molle that it be strung up until the morning. My idea was vetoed by voice vote. In a surprisingly short time the Irianese had hacked the creature into two-inch steaks which were thrown into an open fire for roasting. Within an hour the reptile had been completely consumed.

Saturday, June 4: The rain continued all night and was still falling heavily at dawn as we headed out of camp in the previous day's still-sodden clothes. We were now in one of the wettest parts of the world, caused by the constant collision of warm, moisture-laden air moving up from the coast with cold air flowing down from the mountain tops. Some sections of the foothills average 300 inches of rain per year. During a later visit to the area, we were soaked by an unbelievable 16 inches of rain in only four hours. The figure was probably even higher, because by the time we were able to look at the rain gauge, it was overflowing. Being out in such a downpour is like standing under a waterfall.

The trail today was much worse than yesterday. A continuous series of ups and downs in knee-deep mud, roots, and rotting vegetation left me by eleven o'clock totally exhausted and flat on my back in the soft moss of a swamp. I had so little strength that I couldn't make the last 200 yards to the rest of our group, who were having lunch on a small patch of elevated ground. They sent me a plate of rice and corned beef, and thus revived, I was able to join them. The sun was shining for the first time in 38 hours.

I compared notes with John Bowenkamp and found that he felt fine in contrast to my own weakened condition. He suggested that I was drinking too much water. Thinking back on how I had fallen face first into every stream we had crossed in the past two days—even into the coffee-colored ones flowing out of swampy areas—I knew that he was right. As a novice subjected to extreme physical exertion in a tropical climate, I craved water as an alcoholic might crave strong drink. In this kind of travel, though, water can be enervating and can cause excessive perspiration. It's better to be a little dehydrated. I resolved to refrain from drinking water at every opportunity, but wondered if I would have the willpower to resist those cool mountain streams.

While I rested with my companions on a log in the welcome sunshine, Moses filled my canteen with tea and, after rummaging about in his effects, found some sugar which he added. This proved to be far superior to the swamp water I had been drinking. For the remainder of my time in New Guinea I was able to limit my daily ration of liquids to a few controlled sips of mountain water plus the two-quart capacity of my tea-filled canteen.

At noon John and I started together toward Camp 5, traveling in single file along a track that was often marked by no more than a patch of moss scuffed from a rock or log. Without knowing exactly where it occurred or how, I soon found myself infused with a second wind that greatly improved my trail performance. I had left John behind and was now with three Irianese in the vanguard of the group.

The second wind seemed at least partly psychological, deriving from a system of objective-setting I was beginning to develop. I would select a tree or a rock 200 or 300 yards up the trail, which I would resolve to reach despite aching lungs. Attaining that objective, I would select another one. I further discovered that during rests my revival was accelerated if I propped my feet up in an elevated position. My objectives, especially when nearing the end of what I considered the limit of my ability, were carefully calculated to include a soft bed of moss on the downhill side of a tree on whose trunk I could put my feet.

My Irianese friends were greatly amused with this performance: my awkward advance over terrain they negotiated effortlessly, my gasps for air as I neared my physical limit, and finally my complete collapse in the moss with feet in the air against a tree. This final act was always good for a round of laughs. They would stand upright for the full five minutes I needed to regain my breath and then would politely fall in behind me as I attacked the next objective.

I also found that objective-setting could be a game in which one could resolve to exceed objectives. Before long I was moving uphill for 20 minutes or more, considerably longer than my prior limit. The increase was not attained immediately, but this particular afternoon was the beginning. At 3 P.M., much less drained than at the same time the previous day, I was walking through the remnants of a camp area established earlier in the month by the advance party. This was Camp 5.

In a swift-flowing tributary of the Tsing River near the camp, I took off my clothes, washing the mud and sweat out of each article. When shirt, pants, socks and underwear had been thrown over a large rock to dry, I relaxed luxuriously in a deep pool of clear mountain water. Takim set up his cook camp at the edge of the river and served hot tea after my swim.

It was nearly five o'clock when John arrived and he immediately threw himself on his bedroll. He said that he had suddenly become ill during the afternoon, had lost his lunch and had been retching ever since. Unable to eat supper, he slept fitfully during the night.

Sunday, June 5: I later came to refer to this day as Black Sunday because it was when the Ertsberg Expedition started to come apart at the seams. During the night Takim came down with malaria and was unable to prepare breakfast. Molle made an attempt to fill in as cook, but it was apparent that he, too, was in a weakened physical condition. The ten coastal porters were all suffering from various aches and pains and there was some doubt as to their ability to continue. John felt no better than the evening before and was unable to retain any food. I was fully aware, and John was too, that no one could travel long in this country without energy-building food. We finally broke camp and got off to a ragged start, but we might not have moved at all without the leadership of our young Dutch friend, Jan Ruygrok. The morale of the whole group was at its lowest ebb, but Ruygrok kept things moving and made sure that each man did his appointed job.

As usual, to compensate for my slow pace, I was off in the vanguard with two or three Irianese to lead me. The terrain was becoming increasingly jagged and precipitous. For three hours we climbed through a steep, rocky gorge, often struggling for toeholds on logs or rocks in the midst of waterfalls. It was clearly the most strenuous climbing to date. It was also my introduction to water alpinism, where the only means of advancing through a canyon is along the bottom where the water flows because the banks are too steep for safe footing. While this is a means of progress, it is a very wet one, and in the higher altitudes, a very cold one, too.

Following in the track of the mountain people posed many dangers that were not immediately obvious. They could draw on their instincts and considerable native intelligence in negotiating the ridges and gorges. But we outsiders lacked their uncanny ability to decide instantaneously just when a leap should be made for the next foothold or handhold. I was often forced to pause with a torrent of water cascading over my shoulders for a minute or more while I tried to think out my next move.

At the top of a knife-edged ridge I found a large group of porters who had passed me on the trail. They were reclining in the soft moss where I joined them for a 15-minute rest.

Nearly straight up for the first three hours, the trail now dropped off steeply. In many places I found I could use the transverse surface roots of large trees like the rungs in a ladder; often they were the only way to keep from pitching headlong on the ground. At last the trail leveled off and turned left along a plain. We could hear the roar of the Tsing, which was coursing through a rocky and impenetrable gorge 1,000 feet below us.

At 1100 I decided I had better wait for John. By noon he had still not caught up. I dispatched a packer ahead with a message for Ruygrok asking him to set up an emergency camp at the first available spot. At 1230 John finally staggered up the trail to the level ground where I was waiting. He was so exhausted it didn't seem possible that he could continue. I suggested that we camp here and plan to return to the coast. I was amazed at the vehemence in his reply. He said that he was going on, that to turn back now would be an act of betrayal to Del Flint and the advance group, and that if he could not reach the next camp, he would sleep in the jungle. These remarks may not have been completely rational but they were evidence of courage. I felt ashamed of my prior suggestion of turning back in defeat.

After a short wait Ruygrok joined us. He said an emergency bivouac was being established about one hour's travel along the trail. He was concerned that a part of our group, including Molle and Takim, had already moved beyond the new camp location toward Camp 6. Ruygrok agreed to stay with John and I set off alone toward the bivouac.

Walking along the narrow shelf that served as a trail along the crest of a ridge, I crossed three spectacular landslides, each extending 2,000 feet to the bottom of a gorge where I could see white water of the Tsing. I had seen many landslides in the Andes of South America but none were as steep or as deep. The trail picked its way between rocks and in places was no more than a series of toeholds in soft, crumbling clay. At one point in the middle of a steep drop I was obliged to fit my left toe into a four-inch slot made by Irianese bare feet and swung my leg over a large rock, making accurate contact on the opposite side with my right toe in another small slot in the soft ground. I was only too aware that while crossing those landslides it would have taken only one misstep on a loose rock to send me plunging headlong to oblivion in the deep canyon below.

Two hours after my arrival John stumbled into camp. It was now obvious that he would not be able to proceed. I helped him into the bed which we prepared in advance of his arrival. That evening, as we were having a frugal meal in which John was unable to join, the coastal packers announced that they would not go any further. We were not surprised, for it had been apparent during the previous two days of travel that these men were not strong enough to carry heavy loads even in the foothills. Their aches and pains were our own multiplied many times.

Monday, June 6: At daybreak we had a conference on how to proceed. Our situation was serious. We were about midway between and several days travel from our supply bases at Omowka and Tsinga. We were behind schedule and nearly out of food. We simply could not stand by and wait for John to recover. John agreed that there was no other choice except for him to return to Omowka, where he could be evacuated to Biak. There was risk in such a long journey, probably about ten days, but at least it was in the direction of food and medical help. We advised him to spend today in bed to gain strength for the trip. Though he had not been able to eat for two days, it was hoped that he would be able to retain some oatmeal before the day was over. Ruygrok, the strongest member of our group and the only one who could communicate with the coastal people, agreed to accompany John.

John and I said good-bye, which was not easy for either of us. As I left camp with my customary two guides, Ruygrok and Moses were supervising the repacking of foodstuff and supplies. After a gradual climb through heavy forest, the trail crossed the Kelogong River, which flows into the Tsing, at an elevation of 1,900 feet, and then went up steeply for 1,600 feet without any intervening dips to Camp 6, a small cluster of huts known as Kelangin. Inhabited by 30 mountain tribesmen, it was the village where Moses had been born.

I found Molle, Takim, and some of our packers in the

largest hut. About 12 by 25 feet in area, the hut had been made available to the expedition by the headman of the village. Molle said he had recovered somewhat from his upset condition of two days before. Takim, though, was seriously ill, flat on his back and shaking with chills. He refused to take any food and we did our best to make him comfortable with extra blankets. Our packers were huddled about an open fire on the bare earth floor, poking at sweet potatoes roasting in the coals.

Ruygrok and Moses came up from the lower camp at three o'clock to obtain food and packers. The coastal people who were to return with Ruygrok and John were in such poor physical condition that they would be unable to carry anything, and while that group had adequate supplies, it needed additional packers.

The group going forward had enough packers, but we had to do something about our nearly depleted food supplies. We prepared a list of required food items, which we gave to a runner selected by Moses. He was to travel all night and deliver the list to the two policemen who had been left behind by the advance group to guard our supplies at Tsinga. The police would then dispatch the food back to Camp 7 in the village of Amkiagema, which we expected to reach on Tuesday. In the meantime, we would have to subsist on sweet potatoes exchanged for small amounts of salt. Salt is highly prized in this area since it is obtained only by drying leaves dipped in a local salt spring. While Ruygrok returned to rejoin John, Moses canvassed surrounding villages for packers willing to accompany the group returning to the coast.

After an evening meal of rice and sweet potatoes, Molle and I set up the bicycle-driven radio equipment as a crowd of more than 30 villagers watched with curiosity. Kelangin was on a high hill from which, we had been told, the Carstensz Top was sometimes visible at sunrise, and we thus hoped to be able to talk to our advance group. Once the equipment was set up, we realized that all the natives whom we had taught to peddle the bicycle had remained with Ruygrok in the lower camp. For 30 minutes, we tried in vain to train a new operator. Finally we evolved a procedure which took full advantage of the large crowd of interested observers. Two local boys sat on the bicycle, one on the seat and the other on the handlebars to hold it down; two more were stationed on each side to give the contraption lateral stability; and two more were placed in a squatting position on the sides and shown how to grab a pedal and turn the pedals by hand. In this peculiar fashion a work force of eight was able to deliver from 60 to 90 seconds of continuous power followed by a 20-second break for rest.

At seven o'clock we intercepted a conversation in Dutch between Gus at the Carstenszweide and the expedition's radio operator at Kokonao. Due to the uncertainty of our source of power, I broke in and asked to speak with Del Flint. He came through loud and clear, but there were gaps in his conversation when our boys stopped cranking the generator. I learned that conditions at the Carstenszweide for the first seven days had been unpleasant with only a single hour of sunshine; that the urgency of establishing a camp for protection from the continuous cold rain and fog had prevented Del from making any detailed inspection of the Ertsberg; that he believed it contained not less than three million tons of ore, but that he could not even guess at its copper content.

For the past few days, the thoughts that had been uppermost in my mind were the nature of the trail and my ability to get through. With some trepidation, I asked Del what conditions were like between Kelangin and the Carstensz-weide. "Each day it gets worse and in some places the trail is sheer horror," I heard him reply. "You might be able to cross from Tsinga to Base Camp in four days, and then two or three days to get up here. But I want to warn you right now that the trail on the second day out of Tsinga is a . . ." At that crucial point my power plant crashed to the earth floor of the hut in a tangle of dark-skinned bodies. When order was restored, we were unable to make further radio contact. It was the last conversation I would have with Del for a week. I was left with six days to wonder and worry what he had been trying to warn me about on that section of the trail. By my second day out of Tsinga, I was fully prepared—or at least I hoped I was—for the worst.

Tuesday, June 7: While Molle prepared breakfast of tea and rice, we discussed what to do about Takim. He had ranted and raved through fevers and chills all night and to me it seemed unlikely that he could move over the trail. Kelangin seemed a poor place to recover from sickness. While this entire area of New Guinea was generally unfavorable for convalescence, we felt that Takim would be better cared for at Camp 7 in the village of Amkaiagema to the north. Here there was a jungle school operated by

Father Koot with the help of two Indonesian-speaking villagers. Here also we hoped would be the provisions we had ordered sent back from the Tsinga depot, one day's march to the north.

Before we had finished breakfast, Takim emerged from his tangled bundle of blankets and staggered over to the open fire. He looked gaunt and feverish. Even his eyes were yellow. Yet he was determined to continue. Though he had been ill with malaria for two days and had eaten practically no food, he subsequently passed me on a steep section of the trail later in the day. When I reached Amkaiagema, he was sound asleep in a schoolteacher's hut.

The division of forces and sickness in our midst was beginning to discompose our travel routine. My guide today, as well as the bearer of my knapsack, was to be a boy from Kelangin, named Martinius, who was no more than 15 years old. Low-lying clouds obscured the hoped-for view of the Carstensz glacier to the north, but a mile to the east, across a steep valley, I could clearly see the next native village, known as Buluwaintema. I didn't know it would take more than three hours of difficult and tiring travel to reach that spot.

The trail from our camp at Kelangin dropped off on a steep gradient for 1,400 feet to the Keemagong River and then climbed a precipitous slope for 1,200 feet to Buluwaintema. The crossing of the fast-flowing Keemagong was evidently a routine matter for most of the local residents. It was effected by utilizing a crude log bridge. Two logs 6 inches in diameter extended 30 feet from the south shore to a large rock in the center of the river. Two smaller logs extended about 20 feet from the rock to the north shore.

When I reached the river, I found Martinius sitting on a rock near the river bank with two old men. A flood during the previous night had washed away the connecting logs from the rock to the north shore. When I indicated that I was going to attempt a crossing anyway, there was a loud outcry. From the sign language they employed I gathered they felt that anyone foolish enough to try a crossing would be tumbled down the river and his head split open like a pumpkin, or its New Guinea equivalent. They indicated that the wisest course of action would be to join them on their rock and wait for more people to arrive who could re-establish the bridge.

One of the older men took some of the local equivalent of tobacco out of his pouch and rolled a cigarette. Pointing to me he made the motion of striking a match followed by a loud "pfft." I indicated that I understood his request but with a negative shake of the head I indicated I could not accommodate him. To get in shape for the trip, I had given up my 30-year cigarette habit. (To this day, I have managed to avoid a relapse.) The older man looked at Martinius, who took out his fire-making kit. He whipped a strand of rattan through a forked stick and past shredded bark which ignited in seconds. With long, thick, homemade cigarettes in their mouths the older men leaned over and casually got a light. Martinius snubbed out the smoldering bark and carefully repackaged his implements.

While my companions smoked, I wandered along the river looking for another possible crossing. The Keemagong is a violent river, and as was the case on all crossings of this type, the point which had been selected by the mountain people was the most feasible and usually the only one within a considerable distance. Glancing at the opposite hillside, I noticed someone coming down the trail from Buluwaintema. It was a woman and I was impressed with her agility and grace in negotiating the difficult trail. I naturally assumed she would wait on her side of the river for bridge reconstruction. To my amazement she picked up a round pole for support and started out into the fast-flowing current. The water was soon swirling above her abbreviated grass skirt. After a slight pause in midstream, she lunged for the large rock on which the other bridge abutment rested and with a catlike motion scrambled up its side. In what appeared to be a deliberately casual manner she then walked 30 feet across the narrow logs to the rock on which we were sitting. She indicated that she would accept a cigarette which was prepared and lit for her by one of the older men. The men then proceeded to cross the river without any further exchange of conversation. They had obviously been shamed and embarrassed and it is quite likely that the bridge went unrepaired for many days.

I was the third to cross. In the awkward fashion I had used on many occasions over the previous four days, I straddled the logs and advanced by pulling myself along in leapfrog fashion. My companions looked on incredulously. From experience, though, I had learned that rubber-soled shoes on wet round logs did not provide stability compared to heavily callused bare feet. The rough, yellowed calluses on Irian feet were usually half an inch thick and more durable than shoe leather. To toughen up his feet, Father Koot regularly went without shoes during his periodic visits to mountain villages.

From the edge of the large rock in the center of the river, I eased my body into the river in such a manner that the flow of the current would hold my feet against some rocky obstruction on the bottom. Then, pushing hard on a pole which had been stuck into the river bottom, I lunged for the shallows on the north shore. Here, the current was less violent, and with a clumsy thrashing movement I was able to reach the bank.

My arrival in the village of Buluwaintema must have been well publicized. The entire population of the village seemed to have turned out to greet me. I touched fingers or snapped knuckles with 50 or 60 people, including two or three wrinkled gray-bearded older men who possibly re-sided in this area 48 years ago when A. F. R. Wollaston, a British alpinist and early explorer of the New Guinea highlands, crossed from the east bank of the Otowka on his way to the Carstensz glacier.

Cutting through the middle of this mountain village was a steep ravine bridged by a large log. Rather than publicly demonstrate the crude manner in which white men crossed slippery logs, I walked down into the ravine and struggled with some difficulty and embarrassment to climb the op-posite bank. This odd procedure was observed in polite silence, but after my departure it must have been a topic of conversation and probably considerable speculation as to my motivations.

About an hour's walk north of Buluwaintema I emerged

from heavy forest growth to a rocky point overlooking the Tsing River. Some 800 feet below me were two terraced areas ranging from 50 to 300 feet above the river. This was the village of Amkaiagema, which was Moses' home-town. The terraces had been formed hundreds, perhaps thousands of years ago by landslides which had dammed the river and formed temporary lakes. On the upper terrace were about 50 native huts and on the lower area, adjacent to the river, Father Koot had established his native school, a church, and a small building he occupied on the three or four occasions he visited this region each year.

Moses greeted me and indicated that I was to sleep in Father Koot's house. It was of thatched construction, raised about three feet off the ground and divided into two rooms. Father Koot apparently used one of these rooms as a private chapel and the other, which had a well-con-structed rattan cot, for sleeping. The bedroom had two windows looking south. The windowpanes were sheets of polyethylene, several of which had warped in the sun and fallen out of their frames. Nevertheless, it was certainly the most comfortable abode I had seen for ten days.

After a shave and a bath in a nearby stream, I sat on the front steps of the hut and watched a group of 16 naked children playing soccer with a slightly deflated ball. Father Koot had taught them the game and I was amazed at their skill and agility. Quick as cats, they passed the ball from one forward to another like professionals, and gave it surprisingly hard boots with their bare feet. During the game they kept up a constant chatter in their native tongue. When one of them managed to drive the ball through two sticks serving as uprights at each end of the field, however, everyone shouted "goal!"

At dusk Moses conducted a service in the church, a large barnlike thatched building with logs on the earth floor serving as pews. I was deeply impressed by the simple ceremony: a high-pitched call from Moses, followed by a deep chant of response from 100 or more local villagers. When it was over, Moses led the entire congregation over to my hut for a round of knuckle snapping. All of the women insisted on leaving a contribution of vegetables on the front steps.

Wednesday, June 8: As had been previously planned, we laid over for a day in this pleasant spot to assemble a new group of porters. This gave me an opportunity to photograph the people who had congregated in front of the hut while I spread my sodden gear to dry in the sun. It also gave me a chance to attend to some of my wounds. My hands had about 25 or 30 festering sores, caused from grabbing thorny vines or trees for support while climbing through the brush. I had long since exhausted my supply of Band-Aids, but Molle did an effective repair job with sulfa salve and gauze. He was more concerned with the condition of my feet.

I had been wearing rubber-soled canvas boots with woven rope inserts, which the French had perfected for use in the Sahara. Probably a half size too small in the first place, they had shrunk further from repeated water immersions. My feet had been so squeezed that the nails on the big and little toes of each foot had turned black, indicating that they would soon drop off, and the middle toe on my left foot was badly infected. Molle did an effective lancing operation and I decided to follow Dr. Watson's prescription for infections: Terramycin pills four times a day. Within 48 hours all infections on my hands and feet had disappeared.

In the afternoon Moses set up employment headquarters on the front steps. A large number of villagers squatted on their heels to witness the proceedings. While most appeared more concerned with the task of picking fleas out of each other's hair than joining the expedition, Moses was able to obtain enough recruits. As each new packer was signed on, he selected his load and made adjustments in the rattan strips which served as a carrying harness.

Thursday, June 9: Takim's malaria was worsening and his fever ranged as high as 104 degrees. Last night two men were required to restrain him from smashing his fists into the side of the hut. I decided he would have to remain in Amkaiagema under the care of the Irianese schoolteachers.

From the rough map I had been studying, it was quite clear that I would do far more walking today in reaching our next camp at Jonkogomo or Tsinga than I had done on any previous day's hike. I had made arrangements for an early start and, having discarded my French desert boots, I was now wearing a pair of calf-length green canvas jungle boots made in Hong Kong. As I passed the schoolteachers' hut with Martinius, Molle came out and explained in halting English that the logs on a bridge I would be crossing about one hour's walk north of the village were not in good condition. Moses, with whom I had developed an effective system of sign language, suggested that I try walking across this bridge rather than use my customary technique of straddling the logs and pulling myself along in leapfrog fashion.

The trail dropped off the lower terrace and for nearly an hour followed the rocky left bank of the Tsing. This was my first close look at the wild, savage river I had been viewing from ridges for the past three days on the trail. It drains a large area of the Nassau Mountains, including the eastern side of the Carstensz glacier. Its waters are very turbid, the result of finely powdered ground rock rather than soil erosion. The grinding of rock on rock occurs both in areas of active glaciation and on the river bottom. I frequently heard the deep boom of large boulders being rolled along by the violent flow of water.

After an hour's walk, I came to the Tsing's confluence with the equally violent Nosolonogong where I saw the bridge Molle had warned me about. Approximately 40 feet long, it connected the south shore of the Nosolonogong to a steep and narrow limestone ridge on the other side. The base of the bridge consisted of two round logs about eight inches in diameter. Between the abutments at the ends of the bridge were two thin strands of rattan at waist height. The strands, it was apparent, were not for grasping but for fingertip balance control.

The south abutment on my side of the river was an elaborate affair of poles bound together by rattan and rising about 20 feet above the bank to equal the elevation of the limestone ridge where the bridge met the north shore. While Martinius watched from the ground, I climbed the cross-members between the vertical and inclined poles of the south abutment that served as rungs. From the top of the abutment, I had a panoramic and rather frightening view of the deep chasm below through which flowed a roaring torrent of tumbling white water. I wondered how suitable a bridge constructed by 100-pound people would be for an individual weighing close to 200 pounds. I understood now Moses' crossing advice. My straddle-and-leapfrog technique probably would have subjected the structure to much more stress than walking across. Walking, though, presented its own problems. Because of the 40-foot span, there would likely be much spring in the two parallel logs. Putting my full weight on one side and then the other could easily create such a severe up-and-down oscillation that I could be thrown off balance and into the river. The solution, I decided, was a short shuffle, advancing the left foot a few inches on one log and then the right foot a few inches on the other log. The steps would have to be quick; I would have to avoid the natural temptation, when terrified, to hesitate.

Taking a deep breath, I extended my hands so that the fingers were touching the parallel strands of rattan and started across. As I approached the center of the span, preventing undue oscillation became more and more difficult. Even a cautious shuffle did not prevent the foot on which I put my weight from quickly sinking eight inches below the other foot. Sweat was running off my nose in a steady stream. Sensing that it would be disastrous to allow my eyes to focus on the raging cataract 40 feet below me, I tried to concentrate on the two round logs. "Don't panic, Wilson, don't panic," I heard my voice saying. The voice seemed to have an authoritative tone, and I continued shuffling forward. About three feet from the north landing, the swing between the two logs was reduced enough so that I was finally able to take a long step to solid ground. When I turned around, I saw Martinius on the south abutment crossing himself. He did this apparently not to prepare himself for his passage over the bridge, but to thank the deity for permitting me to make it successfully. He came across on one log in a matter of seconds without even bothering to touch the parallel strands of rattan.

The ease with which Martinius had negotiated the bridge in contrast to my own clumsy and laborious effort clearly showed that the proper technique of New Guinea bridge-crossing requires years of training. Resting in the scrub growth on the limestone ridge, I realized a large part of my available energy for the day had been drained away. I wondered how I would be able to summon strength for the long trail ahead.

My altimeter showed an elevation of 2,500 feet at the river confluence. During the next seven hours the trail followed the limestone ridge to an elevation of 6,100 feet before dropping off into some partially cleared fields. In many places the ridge was only a few feet wide, affording an often spectacular view of the two rivers 2,000 to 2,500 feet below. In the far distance on the east bank of the Tsing, I occasionally saw small clusters of huts. From time to time our packers would stop and cry out in unison in a high-pitched note which they would continue until there was an answering call from across the valley.

Sometime before noon I emerged from scrub trees into

an open grassland, climbed over a series of huge logs set
crosswise to the ridge like a fence and came upon a perfectly flat area on top of a ridge about 40 feet wide and
possibly twice as long. Most of our packers were sitting
here in the warm sunshine. They had passed me all morning as I toiled up the slope and now appeared quite rested
as I flopped down among them. In the Andes of Colombia
and Ecuador I had seen many manmade flat areas at the
crest of steep ridges that were used for dwellings or observation points. It occurred to me that this level spot
probably had some similar significance. (See Appendix
Note 3 for an account of this spot by the leader of an
earlier expedition.)

Dropping off the east flank of the ridge into partially
cleared fields, the trail wandered about in seemingly aimless fashion, probably to accommodate the members of
various local family groups in harvesting sweet potatoes.
Having lost 1,000 feet of elevation, it maddeningly recovered 900 of them in a steep slope ending at our Tsinga
supply base directly above the small village of Jonkogomo.

At dusk, as I neared the end of that last long climb, I
could see, at the top of the hill, the brilliant red, white,
and blue Dutch flag waving from a mast planted at the top
of a dead tree. Looking up, I could also see numerous faces
peering at me from the top of the ridge, including our two
local police constables whom I recognized by their uni-
forms. One was slight and quite dark. The other was almost fair of skin and nearly six feet tall. He was, in fact,
more Polynesian than Melanesian. He grabbed my hand
and pulled me over the top of the ridge into an open-

sided tent which had accommodations for five bedrolls. It was now after six o'clock in the evening. I had been on the trail for 11 hours and was completely done in. Although I was the first to leave Amkaiagema, I was the last by an hour or so to reach Tsinga. Eleven hours on the trail between these two points was not a new speed record, but for Wilson, who on several occasions felt that he would have to camp on the trail, it was quite an achievement.

The two constables pulled off my sodden clothes, gave me a vigorous toweling, inflated my air mattress, spread blankets, and shoved me into bed. Within a few minutes I was sipping hot chicken noodle soup. I had never felt more comfortable or relaxed in my life. At that point I would have been reluctant to have changed places with any man.

Friday, June 10: I awoke with that first dramatic view of the Carstensz Top that I described at the beginning of the book. It seemed a very fitting reward for the previous day's many long hours on the trail. That view had also deeply impressed Wollaston, who reported to the Royal Geographic Society after his first visit to New Guinea:

> We spent 15 months in the Mimika region, and during that time reached a point about halfway to the Snowy Mountains; but I think that even if we had spent twice that time in the country we should not have made much further progress. During those 15 months, we had a view, on I dare say some 100 mornings, of the distant snows of Mount Idenburg and the Carstensz Top, and anyone who has a love of mountains can understand how tantalizing it was, day after

day, to see those virgin peaks so comparably close at hand and yet as unattainable as the mountains of the moon.

We decided that this would be a day of rest and reorganization. We paid off the bearers from Kelangin and assembled a new crew for the push to Base Camp, three days' march to the west. While Moses handed out parangs and other trade goods, I explored the low slopes adjacent to our camp and saw many new species of birds and flowering plants. All could probably have been observed on the trail. When slogging along over rough ground, though, the eyes of the neophyte must remain focused on the point where the next step will be taken. Every time I tried to walk and observe the scenery at the same time, I found myself on the verge of a dangerous fall. In contrast to my plodding gait, the mountain tribesmen on level ground and even on rough downhill stretches of the trail proceeded at a dog trot. With just a glance at the trail, they were able to assess every bump and irregularity for the next 30 or 40 feet, file such information in their heads as one might program a computer, and then trot along with eyes raised and feet landing at just the proper point on each rock or log over the previously scanned stretch of ground.

I decided that Molle should remain at Tsinga with the radio equipment to act as liaison for air drops to supply the return trip to the coast. During the afternoon, Molle set up the transmitter on a solid bamboo platform with two directional aerials, one for Kokonao and one for the Carstenszweide. We found, though, that we could not talk to Gus

and Del from Tsinga, for the Carstenszweide was in a
deep valley whose open end pointed south toward Kokonao. By relay from Kokonao I learned that the last air drop
at the Carstenszweide had been successfully completed on
June 7th and that it included my duffle bag of cold weather
gear which had been responsible for the week's delay at
Biak.

After new porters had been recruited, Moses and I
surveyed our food supplies. For the trail and for the overnight camps en route we selected a stock of miniature
Dutch chocolate bars, dehydrated soups, canned beef, and
sardines. We also filled up a five-gallon oil tin with such
incidentals as rice, oatmeal, condensed milk, tea, and dried
fish from Hong Kong. The following day, the advancing
group would be reduced to a party of ten: Moses, myself,
and eight porters. Three days later, it would be reduced to
nine. Moses, the mainstay of the expedition, would be gone
and I would be the only remaining member of the 50-man
force that had left the coast 16 days earlier.

Chapter Five

Saturday, June 11: That fantastic scene of the ice-covered Carstensz Top was visible again this morning, but again only for a brief interval. Heavy clouds from the south coast of the island completely obscured the mountains above 10,000 feet. I prepared a cable for New York, estimating the date when I expected to join the advance group at the top. As I headed out on the trail at 0700 I could hear Molle on the transmitter repeating the initial phrase over and over again. It usually required about 30 minutes to send a 20-word message in English because neither the sender nor the receiver was fluent in that language. Molle would repeat three-word phrases about six times hoping the receiver would understand the complete phrase, and would spell out each word, using the Dutch or Malay alphabet, whichever suited his fancy.

The trail today involved a drop of 1,100 feet to the Magogong River, a climb of 2,300 feet to the top of the same ridge I had followed for several hours two days previously after crossing the Nosolonogong bridge, and then a drop of 2,000 feet to Camp 9 on a gravel bar on the Nosolonogong's west bank. After the first drop below Tsinga, the trail followed the shallow Magogong up a steep gradient. We had not proceeded far along the trail when Martinius indicated that he was sick and unable to proceed and that he would have to transfer his load to someone else. There was nothing for me to do but sit on a large boulder in the middle of the river and wait for one of the other porters to come along. While I was munching on a chocolate bar, a mountain woman appeared with her young daughter, who was no more than three feet tall. They had a brief conversation with my guide, who indicated that I should leave my load with him and follow the two females.

After I had given them a couple of my chocolate bars, they started up the streambed in their normal swift fashion, leaping from boulder to boulder, and promptly left me panting in the rear. Soon they learned to adjust their pace to mine and when I stopped, they also stopped. After two hours of climbing in the clear, cool mountain air, they left the streambed and scrambled up the north bank. I could see no strong evidence of a trail and wondered if they were tiring of leading me and were trying to lose me in the jungle.

Within 100 yards of the stream we came upon a small level spot covered with pieces of charred wood, indicating that once there had been a dwelling here. Beyond the level spot was a clearing in which sweet potatoes were growing, the apparent objective of my new guides. They sat down and started to eat the tops of a light green plant that appeared to be growing wild around the edge of the level area. I tried some, too, and found the herb to be rather refreshing, with a taste comparable to a mixture of parsley and dill. After a brief rest, mother and daughter started digging sweet potatoes with a wooden implement which must have been hidden on the premises.

From what I later learned of the close family ties among the mountain Irianese, I would judge that one of the ancestors of the woman and her daughter had lived at this spot. The charred pieces of wood may have been the remains of the ancestor's funeral pyre. It was common to cremate elderly family members while they were still alive in funeral pyres near their huts. No one would ever again live on that particular spot, but the field would remain family property. The fact that people were willing to undertake what amounted to a round-trip walk of four hours just to harvest a few vegetables was an indication of the shortage of arable land in the mountains.

It was not long before a sturdy young boy appeared bearing my knapsack. He ate some of the light green herb and then pointed to the northwest indicating that we should be on our way. Returning to the stream bed after waving good-bye to the woman and her daughter, we followed it to its source in a series of springs in a swampy area near the top of a very narrow ridge covered with giant tree ferns and other types of flora I had never seen before. The last 300 feet to the top of the ridge was a sharp sandstone escarpment along which I slowly crawled in crablike fashion.

We paused at the top of the ridge, and I distributed

some of my chocolate bars to the packers. Like other mountain tribesmen to whom I had given the bars, they were not sure at first what to make of the objects. Removing the aluminum foil wrapper of one bar, I showed them that it should be eaten. They were more interested in the wrapper than the contents. After consuming the chocolate they carefully stowed the wrappers in their net bags.

For the past several days, ever since leaving the village of Buluwaintema, we had been following the route of a 1912 expedition led by A. F. R. Wollaston, which was mostly along well-worn native trails. From where we were now standing, Wollaston had continued north along the ridge on what was seemingly an easier and more direct track to the Carstensz Top. This, though, had been his fatal mistake, for the ridge eventually ends, as I noted earlier, in a near vertical and unscalable wall many hundreds of feet high. Having gotten within just a few miles of the peak, he was forced to turn around and go back.

Instead of proceeding north, our track left the ridge and headed west along a trail that was entirely the creation of our advance group, since apparently there had never been any regular travel between the regions around the villages of Jonkogomo near Tsinga and Wa, just south of Base Camp.

The initial portion of the new trail was not heartening. Ridge tops in the Nassau Mountains tend to drop off sharply, a characteristic that later made road construction along them very hazardous. The trail led immediately down an abrupt incline that had to be negotiated by swinging on vines guided only occasionally by footholds in the rock. On somewhat more level ground, we continued descending through trees heavily coated with moss to a small stream which we followed downhill in the reverse procedure of the morning climb. For the last hour of the day's journey, we picked our way through a vast rockfall whose origin had been in a deep canyon south of our track.

It was a major earth movement and must have been very recent. During the past two or three years several million tons of rock had broken loose from a mountainside and picking up mud and water had flowed downhill like lava for several miles, leaving behind acres of huge boulders. The landslide had stopped in the Nosolonogong River, damming it at that point for a sufficient length of time to leave a sizable bar of gravel. As I picked my way down through the rocks, I could see the ridgepoles of a former camp on the west side of that gravel bar. This was Camp 9 established by the advance party during early May. I later learned that Del Flint and his party had never occupied the camp because a sudden rise of the river drove them to higher ground.

I crossed the Nosolonogong for the second time some 3,000 feet in elevation above that frightening bridge crossing. The river was now shrunken to a benign and fordable stream. I waded through clear mountain water flowing hip deep over rounded river boulders. After a swim and a shave, I started a fire, and before the first of the bearers arrived I had set up the tent. Tea was brewing by the time Moses came into camp with the last two packers. While we were having our third cup of hot tea in the late afternoon, three Irianese came down the west ridge and trotted over to our fire. They were armed to the teeth with bows and hunting arrows. Each was carrying a silk parachute from the air drop at Base Camp. Moses arranged for the three silk parachutes to be cut up and divided among a group of 10 packers who served the advance party on the trail between Tsinga and Base Camp. I learned from Moses that the men had made the distance from Base Camp in a little more than eight hours, but that did not stop me from wondering about my ability to retrace their steps in two hard days of travel.

Sunday, June 12: This was the "second day out of Tsinga" for which I had been mentally preparing myself ever since my brief and interrupted conversation with Del Flint on June 6th at Kelangin when my power plant collapsed.

I had given considerable thought to what Del really wanted to tell me about the trail. I felt it would be either a very long day, say 13 or 14 hours on the trail, or a longer than average day, say 10 hours, with even more challenges than had been encountered previously. The latter seemed most likely. To pace myself I made a mental note that each half hour would represent 5% of the day's journey. Though one might be damned tired after six hours of travel one would have a strong compulsion to push on, because only 40% of the hike remained to be completed. Without this mental preparation, I reasoned, the mind after six hours would console the body by telling the body it was being subjected to undue strain; the body would feel sorry for itself; and within a short time there would be a complete collapse. Such a stalled human mechanism is not easily stimulated to further effort.

I left Camp 9 at dawn and during the first half hour probably glanced at my watch 20 times. The 5% completion point was effected at the confluence of the Nosolonogong with a stream that seemed to be flowing from the general direction of a high mountain known as the Zaagkam. The trail turned west along the smaller stream, crossing back and forth through shallow rapids and a few deep pools. While the water was seldom above my knees, I frequently had to lend a hand to some of the packers struggling in water over their thighs.

Reaching the bottom of a high, steep canyon, I got my first view of what Del Flint had in mind. From at least 2,000 feet above me, the stream cascaded down the canyon toward me in an almost continuous series of waterfalls. Several had aggregate drops of 200 feet or more and one had a spectacular plunge of about 1,000 feet made up of a sequence of individual 20- to 100-foot falls. I could see the first contingent of packers already at the midpoint of the canyon climbing across a bare patch of rock. They seemed glued to the cliff and I had the strange impression they were sliding up the rock wall while the white water rushed past them in the other direction.

For about three hours, I ascended at a sharp angle along the wet, slippery rocks on the left bank of the stream. To my right, the rocks sloped abruptly toward the deep groove in which the stream was flowing. At times, the slope toward the roaring waterfalls, sometimes hundreds of feet below me, was so sheer that I had to pull my way along with a tangled mat of roots and trailing vines which grew at the edge of the riverbed.

When the trail finally reached the top of the cascades,

my watch indicated that I had completed 45% of my expected day's travel. The trail now continued upward along the streambed at a much more gradual incline. For another three hours, in a steady rain, I plodded along slowly with an occasional break and at a pace which I believed would allow me to reach camp before dark. It was with much surprise and joy when, after completing only 65% of my assumed journey, I looked up from the stream bottom and saw the forward group of packers trying to pull a tarpaulin over some poles at the top of the bank. I had reached Camp 10. It had been the easiest day of travel to date, not because it was less difficult than any of the preceding hikes, but because I had prepared myself mentally for something far worse.

Del Flint later told me climbing up that canyon had been the worst day of his life. Not only had I been prepared for it but I had the advantage of some rudimentary ladders left by the advance group. Del had only vines and roots, several of which broke under the strain. On a number of frightening occasions, he came close to falling into the bottom of the canyon. He didn't think he was going to make it. On the trip back down, his experience would become such a psychological block that I wasn't sure he would be able to set foot in the canyon again.

Camp 10 was at an elevation of 8,700 feet, and by the time we had constructed our crude shelter and spread our bedrolls in the cold rain I was shaking with chills. Even after I had changed into dry clothes, there was a considerable wait for warmth. The only available fuel was wet, moss-covered branches which eventually ignited only because of the considerable fire-making skill of the Irianese.

Monday, June 13: The wind whistled through our tent during the night driving a fine rain before it. One thin woolen blanket was not adequate in the damp cold and sound sleep was impossible. At five o'clock I heard the packers cutting wood chips and an hour later the pungent odor of smoke filtered under the tarpaulin. But I remained under my thin blanket until 0630 when the first rays of the sun struck a high ridge leading to the southeast flank of the Carstensz Top.

The morning was bright and clear. Beautifully framed in the open end of our tarpaulin tent was the Zaagkam, a magnificent mountain no more than a mile or two away, rising up to 14,500 feet in a steep pyramidal summit resembling the Matterhorn in Switzerland. This was the same mountain that Wollaston had viewed in 1912 from the opposite direction and about six miles northeast of our Camp 10. In describing the scene as he emerged from scrub growth into open rocky ground above the tree line, Wollaston wrote, "To the west the view was limited by the splendid mass of an unnamed mountain, a buttress of Mount Carstensz, resplendent with bands of yellow and red rock." The yellow and red bands which he noted are due to the oxidation of iron pyrite with which much of the limestone and quartzite mass is impregnated.

I took several photographs of the Zaagkam in the early morning light and then started up a steep trail. The character of the vegetation changed abruptly. The trees were now all heavily coated with moss. As I neared the top of the ridge, there was little more than scrub growth. One strange variety of tree had no leaves. The ends of its branches were covered with dark green plates, resembling the ears of a prickly pear cactus, and about one-half inch thick and three or four inches in diameter.

The view from the top of the ridge—really the junction of three ridges at an elevation of 9,600 feet—was fantastic. From the time we awakened, the sky had remained absolutely cloudless. It was the longest continuous stretch of clear weather I ever observed in New Guinea. Even the Irianese were impressed. Standing on the ridge, they pointed out to each other distant features with which they were familiar.

To the southeast and east, we could see the long series of ridges we had followed during the six days of travel from Belakmakema to Tsinga. To the north the Zaagkam blocked out the central portion of the Carstensz glacier, but the eastern end of the ice field and the snow-covered top of Mt. Idenburg to the northwest glistened brilliantly in the sunlight. The scar of a large landslide above Dozy's Base Camp, which Van Gruisen had noted during the reconnaissance flight, stood out prominently to the west and gave me a measure of the distance remaining to our base camp eight miles to the west. To the southwest and south, extending to the Arafura Sea, was a vast panorama of jungle and swamp cut by numerous rivers with which I had become familiar during my previous six months of studying maps and planning various approaches from the south coast to the Ertsberg. The configuration of rivers was particularly clear just below the foothills where great volumes of water pour out of steep-walled canyons onto relatively flat ground, cutting wide sinuous swaths through the jungle growth and depositing prominent bars of sand and gravel.

A sharp bend to the east of a particularly wide series of gravel bars identified the Otowka. The Minajerwi was easily recognized by its three large tributaries converging about midway between the foothills and the seacoast; a long section of the Dozy route could be followed from the junction of the Ajkwa and Otomona Rivers to the confluence of the east and west branches of the Otomona and the full length of what was called the Kemaboe Ridge, leading to his base camp at the village of Wa. The 1936 expedition had required five days of hard toil to cut its way through six miles of tangled roots and vines on the Kemaboe Ridge.

From our observation point the trail gradually descended to the northwest, passing first through two strange upland swamps consisting of grass hummocks growing in soft black muck, and then dropping off rapidly through thickets of bamboo and rattan. After three hours of difficult walking, we encountered the headwaters of the Oetekenogong at a level area where a recent rock slide from several thousand feet up the southeast face of the Zaagkam had come to rest. It was a good place to pause and relax leg and thigh muscles unused to several hours of steady downhill walking. Though covered by boulders and crushed fragments of limestone, the spot was unusually flat and open. Del and I later selected it as the site of the town that would have to be built for the work force once development of the Ertsberg began. The town, named Tembagapura ("Coppertown"), now has 3,000 residents.

The trail now followed the Oetekenogong, crossing frequently from one bank to the other to take advantage of the terrain. My second pair of jungle boots had started to come apart while descending the ridge and after a few river crossings the disintegration was nearly complete. Sand and small pebbles sluiced in and out of open holes while I walked in the shallow water at the river edge, and I had to be careful that the flopping soles did not trip me in the rocks along the river bank.

As we rounded each bend in the river, I expected to have a view of Base Camp. Finally, after an hour and a half of travel from the rock fall, my guides pointed out a group of people waving to us from the top of the bank a quarter of a mile downstream. The welcoming committee which had come from Base Camp to greet me included Vincent (Bob) Croin, the Dutch police officer assigned to our expedition; Waromi, one of our two camp foremen; and a group of mountain people from Wa.

I followed the others downstream, scrambling over large rocks and splashing through shallow water in my ridiculous flopping footgear. Within a few minutes we were at Base Camp just a mile above sea level and 4,400 feet below our morning observation point.

Base Camp was a curious assortment of buildings and shelters located on a narrow ridge of stream-deposited boulders and low vegetation at the junction of the Oetekenogong, which I had been following downstream for the past two hours, and the Aghawagong, which drained the southwest flank of the Carstensz glacier. I was amazed at what Bob and his local helpers had been able to do in less than three weeks since supplies and equipment had been air-dropped into this Stone Age world.

Near Base Camp's main building, from which a brilliant red, white, and blue Dutch flag was waving gaily in the breeze, a small group of stocky, dark-skinned and heavily bearded men, clad in blue shirts, shorts, and tin earrings, were arranging long poles and strands of rattan which they had dragged into camp from the adjoining forest. They were part of the 19-man contingent of porters from the town of Sansapor on the west coast of New Guinea who had accompanied the advance party from Sorong. They were supposed to be on a two-day rest period from their principal duty of shuttling supplies from Base Camp to the Ertsberg, but instead, as a group of local women and children watched, they were assembling the components of a bridge which was to be constructed the following day across the Aghawagong River, a narrow but treacherous torrent of white water surging around large boulders. The bridge would replace the present crossing, a precarious series of thin poles half-buried in a waterfall.

The Irianese from Sansapor are racially very different from the Irianese in southern New Guinea. They are tall and rugged. Culturally and intellectually, they are much more advanced than the local natives, whose primitive ways they regard with some amusement and to whom they feel very superior. I was struck most, though, by their industriousness and initiative. It seemed that wherever they went, they improved the conditions around them. Earlier, after they had arrived in Omowka, some Sansaporese noticed how the coastal villagers struggled every day to climb from their canoes in the river up a slippery, muddy, 20-foot high bank. In three days, the Sansaporese built a beautiful set of wooden steps. Nobody asked them to do it. They just did it.

Close to where the Sansaporese were working on the bridge was a cook tent made from a mottled yellow and green jungle-camouflaged silk parachute which had been formed into a large cone by a center pole. The lower edge was tied down to tent pegs with strands of rattan so as to leave a four-foot air space above ground. It covered three open fires on which pots of rice and other foodstuffs were constantly being prepared. Next to the cook camp was the "godown," a long narrow structure containing all of the supplies delivered during the eight air drops in late May. The roof of this building was made from several large plastic-coated nylon tarpaulins thrown over a long ridgepole, and the siding consisted of burlap sacks sewn together. It was certainly the best-stocked grocery store between Biak and Australia. We sometimes had the feeling, though, that our supplier, the NNG Trading Company otherwise known as Nigimy, had attempted to unload its slow-moving items on our expedition. The international flavor of a trading company doing business in the South Pacific was clearly evident from the labels on the neat rows of cans and packages on the crude shelves. There was rice from Siam, canned beef from Argentina, sardines from Norway, dried fish from Hong Kong, canned fruit from Japan, crabmeat from Russia, dehydrated soups from Holland, dehydrated vegetables from America, and canned peanuts from Jamaica.

Quarters for sleeping and eating were located at the southern end of camp. The building had been constructed with a light framework of round poles covered with nylon tarpaulins on the roof and with burlap sacking on the sides. At several points along the walls were hinged sections which could be pushed outward and held in the open position by short sticks to form windows for ventilation. The outstanding feature of the structure was the spacious picture window nearly surrounding the area reserved for eating. It had been formed by a three-foot sheet of polyethylene wrapped around a section of the east and west sides and the entire southern end of the building, giving an unobstructed view of the two rivers which converged 100 feet below camp and the steep canyon through which the east branch of the Otomona flowed to the southwest. As a group of curious Irianese crowded about the open windows of the sleeping area to watch, I changed into dry clothing and tried out one of the four canvas cots. It was the most comfortable bed I had felt since leaving Biak 15 days before. Bob Croin woke me up at sunset and said it was time to break the seal on a bottle of Scotch whiskey which the Kroonduif pilots had dropped as a gift on their last pass over Base Camp on May 24th. Scotch whiskey mixed with water from melting glaciers is an excellent drink at any time. The fact that our bottle was probably the only supply within 15 days of travel seemed to enhance its taste even further. Moses and the rest of our bearers had arrived in the late afternoon and it was decided that we should have a day of rest before starting what we hoped would be the final two-day climb to the Ertsberg.

Tuesday, June 14: This morning I watched the Sansaporese build their bridge. Waromi was the foreman and two of the Sansaporese acted as his helpers. The actual labor of construction was done by the mountain tribesmen. Work was started at dawn. Abutments about 10 feet high and made of round poles woven together with rattan were first erected on each bank. A rope was thrown from the west bank and was attached to an 8-inch-wide log about 40 feet long. With 10 men tugging from the west bank it was pulled into position and firmly anchored to each abutment with tight wrappings of rattan. This maneuver was repeated with two more logs, and the three long logs were tightly laced together at three-foot intervals to form the bridge platform. Up to this point all lacing and anchoring had been done with rattan which the workers tore into strands with their teeth and then tied into knots of their own design. But at our insistence, and as a concession to the modern age, they stretched two parallel strands of ⅜-inch nylon rope between the abutments for handholds, which were tied to the platform. By noon the bridge was completed.

There was no official inauguration ceremony. Those who had worked on the job gathered at the front door of our living quarters where each was paid by Bob Croin with three strands of colored beads. The initial cost of 100 strands of these beads must have been no more than a dollar. They had been transported, though, from Czechoslovakia to Hong Kong and then on to Biak where they had been loaded on a plane and finally air-dropped into the jungle. Their delivered cost was probably equal to their weight in gold.

The first pedestrian to use the bridge was a small boy from the village of Wa with a heavy load of sweet potatoes thrown over his shoulder in a net bag. He crossed in a casual manner as though the bridge had always been there.

The main chore of the day accomplished, we devoted the afternoon to bathing in the Oetekenogong. Though chilly, it was about 10 degrees warmer than the water from the melting glacier in the adjoining Aghawagong. The superiority of the Oetekenogong for swimming had clearly been the prime consideration in the location of a bamboo and rattan structure that served as a self-flushing latrine. It overhung the bank of the Aghawagong.

Wednesday, June 15: During the night Moses was hit with a severe attack of fever. By morning it was obvious that the little man who had led both sections of our expedition from the south coast, and who had been my loyal companion and friend on the rugged trail west from Tsinga, could not continue. There was no common language in which we could converse when I visited him at his tent, but his eyes clearly showed disappointment in not being able to accompany me to the Carstenszweide. My new companion would be Bob Croin, who had been waiting for an excuse to make the trip to the top.

At 0710, old man Wilson, the last active member of the expedition's main body, crossed the new Aghawagong bridge with two Sansaporese guides and headed up the trail. Though he gave us an hour head start, Bob Croin, a vigorous young man, passed us during the third hour. Thereafter the trail was well marked by bits of aluminum foil from the chocolate bars he consumed en route.

The ten-mile stretch from Base Camp to the Ertsberg was extraordinarily rugged. It had nearly defeated the Dozy Expedition. Largely from wandering into numerous dead-end ravines, Dozy spent two weeks reaching the southern edge of the Carstenszweide. Certain that no trace of the Dozy trail would remain after 24 years, the advance group had planned to spend at least as much time establishing a new track. Having no particular reason to travel back and forth to the Carstenszweide, the mountain tribesmen would certainly not have maintained a trail of their own.

Fortunately, while the advance party was constructing Base Camp, Moses learned that the headman of the village of Wa, close to where Dozy's base camp had been situated, had once walked over the Dozy trail with his father when he was a boy. He agreed to serve as a guide. His memory of the route proved infallible. He frequently had to climb tall trees to get a better look at the landscape ahead, but he never once made a false move up a blind canyon. With his help, the advance party made it to the Carstenszweide in only five days. With the trail now well established, I hoped to do it in two.

From the Aghawagong bridge, the trail followed the west bank of the river and then veered away to climb a steep slope terminating at 6,000 feet in a glacial moraine, a level, grassy area covered by rocky debris, mainly gravel and limestone remnants, that had been pushed in front of the advancing Carstensz glacier during the Pleistocene epoch, more commonly known as the Ice Age, which began nearly two million years ago. So far as I can determine, this is the lowest elevation at which there remains any

record of active glaciers during this epoch at a latitude—
4 degrees, 30 minutes—so close to the equator. The flat top
of the moraine was an upland swamp harboring a curious
assortment of tree ferns and other tropical plants.

The trail descended several hundred feet, crossed over
a fallen log to the east bank of the river, and climbed back
near the top of a high cliff on the other side of the river.
With the river boiling down a canyon 200 feet below, we
walked cautiously along a narrow shelf. Across the canyon
a large tributary of the Aghawagong tumbled out of a
valley and made a spectacular dive of nearly a thousand
feet. Near the bottom it appeared that a major part of the
water flow was converted to spray and mist which obscured
much of the dense vegetation near the confluence of the
two streams.

After a gradual climb to 6,900 feet, the trail reached the
foot of another moraine and climbed steeply. The climb
was relieved somewhat by a series of crude ladders erected
by the Sansaporese during their shuttle service of the
previous two weeks. The top of the moraine, an elevation
of 7,800 feet, was another upland swampy area of poor
drainage. The tropical tree ferns of the lower moraine were
replaced by other strange forms of vegetation covered with
a deep layer of moss. We came upon a native hut reported
by Dozy 24 years earlier. It was uninhabited, but remarkably well preserved.

After climbing to 8,400 feet, the trail dropped off steeply
for 400 feet to Camp 12 on the east bank of the river. This
spot had first been occupied by the Dozy Expedition and
probably its principal attraction was a natural cave formed

by the overhang of a large slab of limestone in a cemented matrix of smaller stones.

Rain was falling as I came puffing into the camp. Having arrived several hours earlier, Bob Croin was relaxing on his bedroll. I spread my own bedroll, which now included two heavy woolen blankets obtained from the Base Camp godown. We were now a few hundred feet lower in elevation than Camp 10, where I had shivered most of the night under a thin cotton cover. The spray from the near-freezing water of the Aghawagong, however, more than offset any increase in the temperature gradient attributable to a drop in altitude. Two mountain tribesmen and several of the Sansaporese porters huddled together for warmth in the smoke-filled cave where the cook camp had been established. Yet despite the roar of the river and the spray, which covered everything with fine beads of water, I slept soundly for 10 hours.

Thursday, June 16: I emerged this morning from a damp bundle of woolen blankets, put on my still-soggy boots and clothes, consumed a bowl of porridge and a large mug of tea, and got myself ready for what I hoped would be the last day of the grueling trek from the coast. I had somehow managed to survive intact the struggle up, down, through, and over innumerable swamps, rivers, ridges, waterfalls, and gorges, but today I had to face perhaps the most formidable obstacle of all: a sheer headwall that rose over 2,000 feet from a canyon bottom to the Carstenszweide meadow where the Ertsberg was located. I remembered well Dozy's description of the troubles his expedition had in conquering it. "It was the toughest individual climb of our entire trip," he had told me.

I took a last sip of tea and started up the trail with two Sansaporese. It followed the bottom of a deep canyon carved out by the Aghawagong. Ragged quartzite ridges soaring up almost vertically four to six thousand feet on either side of us were outlined sharply against a narrow aperture of blue sky. The direct rays of the sun, which had risen behind the 16,500-foot Carstensz Top, would not reach the canyon bottom until noon.

For three hours, the gradually ascending track crossed and recrossed the river. We jumped from rock to rock trying to avoid unnecessary immersion in the icy water.

I came upon the headwall at about ten o'clock after walking around a sharp bend in the canyon. The Aghawagong and one of its tributaries coursed down the dark slope in a series of breathtaking waterfalls. I had prepared myself for the coming ordeal, but I must confess I suffered a pang of dismay when I first looked up at that savage obstacle.

I unwrapped and ate two small chocolate bars, took a long drink of sweetened tea from my canteen, and proceeded with as much resolve as I could muster. The trail angled across a slope of barren igneous rock made passable by a series of cracks and grooves. It then led almost straight up a cliff broken from time to time by narrow shelves. The steep incline was covered with deep moss, vines, and scrub growth which made climbing both arduous and painful. To advance, I first had to push with feet and knees and then desperately pull with hands and elbows. Hips and shoulders were required to coordinate the effort. Soon it seemed like every part of my anatomy was expressing an aching protest. Though the steepness of the slope eventually diminished, the depth of the moss and muck increased.

Progress came to consist of two or three steps forward, followed by at least one slithering slide backward.

Two endless hours later, I clawed through a final dense tangle of roots and vines. Trying to catch my breath, I stood on the relatively level ground of the Carstenszweide, a boulder-strewn valley at an elevation of 11,600 feet. Bob Croin, who was about 100 feet ahead of me and had been waiting for me, fired two rounds from his police revolver. The sharp sounds echoed and re-echoed from side to side of the narrow valley. Del Flint and Gus Wintraecken, cutting samples with a crew of Sansaporese on the fog-shrouded top of the Ertsberg, were advised of our arrival. From somewhere in the clouds a greeting floated down to us, "Welcome to the Ertsberg." A voice from above seemed so appropriate in the cathedral-like atmosphere of this place that neither Bob nor I thought of returning the greeting.

About a hundred yards away was the object of all my exertions for the past 17 days. Jet black and glistening wet from the steady rain, the west wall of the Ertsberg rose up several hundred feet until it was lost in a layer of mist. The ore mountain dominated the landscape. It was much larger than I had imagined in numerous prior attempts to visualize it.

I stared at it for several minutes before slowly, almost reverently, approaching its base. I had known about the Ertsberg for less than a year, but I felt like I had been waiting for this moment all my life. With my geology pick, I struck several of the boulders on the ground that had broken off the main mass. When the thin black outside layer—the product of oxidation—was chipped away, I saw on every piece the gleaming golden color of chalcopyrite, a sulphide of iron and copper. Everything Dozy had told me at our meeting at The Hague seemed confirmed.

I walked past the southeast corner of the Ertsberg and over a boggy patch of ground to a small gully through which an icy stream flowed. On the opposite side was a small grassy area that had been named Bosseweide in honor of Pieter van Bosse, a director of the East Borneo Company who had been associated with it since 1913. On Bosseweide was our camp, which Del Flint, more informally, had christened the "Stone Age Hilton." It consisted of a cook tent, fashioned from a silk parachute formed into a cone similar to the one at Base Camp, a long, partially open godown tent, a series of pup tents for the resident Sansaporese, and a rather odd-looking sleeping tent of billowing silk. It had a double layer of parachute cloth on all sides and a canopy fashioned from a mottled green and yellow parachute that extended eight feet out from the entrance. Flying from 15-foot poles were Dutch and American flags, which had been unfurled in honor of our arrival and continued to fly day and night during our stay.

As I walked across the parade ground in front of the flags I was greeted by Mardjuki, the Indo-Chinese cook, and his faithful helpers Tsumanboi, Jove, and Makamor, who stood at attention in front of the cook camp. Turning into the sleeping tent, I tripped over some folds of silk cloth and fell flat on my face on a sheet of white canvas. In this ignominious manner I ended my climb to the Ertsberg.

The numerous folds of silk parachute cloth which had tripped me also provided protection from the curious stares of the local staff who had been on hand to welcome me. I was therefore able without undue embarrassment to remain for several moments in a prone position from which I had a ground level view of what would be home for the next few weeks.

The tent had been ingeniously designed to suit living conditions near the Ertsberg. Even with the sun shining, I later found out, the daily temperature rarely rose above the forties. During the inevitable afternoon rain and fog, it dropped into the thirties. At night, it was frequently below freezing.* It was thus necessary to have a tent where one could not only sleep but change clothes, eat, and relax. To protect against the weather, the tent had a plastic-coated nylon tarpaulin as a roof and silk parachutes salvaged from air drops as side walls. Because the ground was soft muck, the floor had three layers: six inches of sawdust that had been used as packing in the air-dropped steel drums covered by heavy-duty polyethylene sheets that in turn were covered by empty burlap bags. A system of ditches had been cut along the sides of all the tents and connected to a main drainage canal running through the camp area. A small river of water flowed continuously down the canal.

Looking toward the northwest and southwest corners of the sleeping tent as I lay on the floor, I could see camp cots covered with sleeping bags, clearly those of Del and Gus.

* In recent years, the climate around the Ertsberg has moderated significantly. The average temperature is five to ten degrees warmer, which has caused the Carstensz glacier to retreat several thousand feet.

In the northwest corner was an unassembled folding camp cot and two large green duffel bags still covered with airline bills of lading and other insignia of international travel. Having been parachuted to the Carstenszweide on June 6th, they were now placed in what I assumed would be my section of the tent.

As I stripped off my wet clothing, I noticed the southeast corner of the tent, which could best be described as a recreation area. There were three camp chairs and a makeshift table on which I saw a deck of cards, a sheet of paper with columns of figures, which I took to be gin-rummy scores, a can of peanuts, three small glasses, and a nearly full bottle of Remy Martin cognac. I was about to sample the contents of the bottle when I noticed on the rear of the table a package of letters neatly tied with a string. The letter on top had a May 19, 1960, dateline from Rowayton, Connecticut, and was addressed to me, Biak, Netherlands New Guinea, c/o Kroonduif for next air drop to the Ertsberg Expedition. It represented an item from the first and possibly last airmail delivery to the Carstenszweide by parachute, and a philatelist or a collector of first covers would probably have made some immediate calculation of its value. To me it was the first news from home since my departure from New York more than a month ago. While reading my wife's letter, I heard a noise outside the tent. Someone in a loud voice said, "Who is drinking our whiskey?"

The head of a heavily bearded man appeared through the folds of silk cloth. Del Flint had accompanied me on numerous mineral explorations in such widely separated places as Cuba, Sicily, Spain, and Poland, but I hardly recognized him. I subsequently learned that Gus had given him the name of "Molotov," but "Abominable Snowman" would have been more appropriate.

In a few minutes Gus joined the reunion. For hours we exchanged individual experiences on the trail, discussed the Ertsberg, sampled cognac and dined on Mardjuki's lunch of Dutch pea soup and an Indonesian dish of spiced fried rice known as *nasi goreng*. Gus and Del were concerned about John Bowenkamp and said that Kokonao on the previous evening had been unable to report any word of his progress since the message I had sent from Camp 5 on the 6th of June.

My two companions helped me assemble my camp cot. With a key from a packsack, I opened my duffle bag and removed a sleeping bag, an inflatable air mattress, an assortment of heavy clothing for cold weather, three pairs of boots, a Fahrenheit thermometer, steel tapes and miscellaneous engineering equipment, crampons or creepers for walking on ice, and a small medical kit. John Bowenkamp's duffel bag contained a duplicate set of equipment, most of which was divided up between Del and Gus.

The thermometer was hung on the flag pole outside our tent where it probably continued to record daily temperature variations for several years after we abandoned our camp. One of the empty duffel bags was appropriated by Mardjuki. When I last saw him several weeks later on the south coast, he had it slung over his back. It probably contained all of his worldly goods.

At 1800 we put on rainsuits and walked up to the radio

shack, about 300 feet west and slightly above our camp and 100 feet from the vertical south wall of the Ertsberg. It was a crude structure, protected from above by a small tarpaulin but open to the wind and the wind-driven rain on all four sides. It faced due south down the steep-walled canyon of the Aghawagong. On an unusually clear day it was possible from this point to see a small section of the Arafura Sea more than 60 miles away.

From the first day of their arrival at Carstenszweide, Gus and Del had established a procedure of calling Kokonao every day at 0700 and 1800. Usually there was no more than a brief exchange between the stations, confirming that all was well, but this evening we hoped to relay cables to New York and The Hague reporting my arrival and supplying first impressions of the ore deposit. While Del pedaled furiously on the bicycle-driven generator, Gus opened the circuit. First in Dutch and then in Malay he started calling the English equivalent of "Carstensz one calling Kokonao, over." After 10 minutes of calling, he gave up, much to the relief of Del Flint on the power plant. A storm between us and the coast had apparently cut us off from the outside world.

Chapter Six

---◆---

Friday, June 17: Thick clouds and rain still enveloped us this morning and we were again unable to raise Kokonao. Protected by an oiled silk rainsuit, I left the tent for an eagerly awaited tour of the Ertsberg with Del Flint. It was a nice way to celebrate my twentieth wedding anniversary.

As I walked around the base of the deposit, I speculated on its geological history. It seemed relatively easy to reconstruct. At one time New Guinea lay beneath the sea, covered by extensive deposits of sedimentary materials settling to the ocean floor. Beginning probably millions of years ago, the central part of the island was thrust upward through the collision of two of the earth's plates, huge sections of crust whose movements shaped and continue to shape our continents, seas, and other surface features. Plate collision is believed to have been responsible for most of the earth's major mountain ranges, such as the Rockies, Andes, and Himalayas. Despite their present lofty elevation, the Nassau Mountains still reflect their ancient past. They are covered with a 1,200-foot-thick layer of limestone, the most common sedimentary material, in which one can still find fossils of marine life. The New Guinea lowlands were created from material washed down from the mountains by many thousands of years of heavy rainfall. The large volume of rain and the consequential rapid rate of erosion was responsible for the unparalleled cragginess of the landscape in the central highlands.

From time to time, the New Guinea mountains, like many other parts of the world along the edge of plate interaction, were subjected to extreme pressure from the earth's magma, the core of molten rock deep below the surface that in its most dramatic form is responsible for volcanoes. This molten rock pushed upward through weaknesses such as faults in the limestone deposits that had been created by the wrenching process of uplift. The limestone was intruded or invaded in places by deposits of igneous rock accompanied by mineralized gases and liquids from the earth's interior. After further faulting, these mineralized solutions chemically dissolved parts of the igneous rock and caused those parts to be replaced by minerals of iron, copper, silver, and gold.

The Ertsberg is an unusually large igneous intrusion with an unusually high degree of mineralization. Most intrusions have only traces of ore. The Ertsberg turned out to be 40% to 50% iron (mainly in the form of magnetite or iron oxide) and 3% copper (mainly chalcopyrite and bornite, both sulphides of iron and copper). Three percent is quite rich for a deposit of copper, which is a much rarer mineral than iron. The Ertsberg also contains certain amounts of even more rare silver and gold.

The Ertsberg's unique and dramatic exposure above

ground was the consequence of glaciation, though that is not the sort of erosion one expects to find so near the equator. Originally, the Ertsberg was completely embedded in a complex of limestone and a variety of chemically and physically altered igneous rocks. Beginning in the early part of the Ice Age and continuing as recently as 5,000 to 10,000 years ago, glaciers moved down from the top of the Carstensz range, slowly cutting their way through the soft limestone and altered rocks and carving out valleys like the Carstenszweide. Being composed of hard ore, the Ertsberg resisted the glaciers and the ice flowed around and over the deposit. As the softer rocks surrounding it were stripped away, the Ertsberg was left standing high above the surrounding terrain. It was later determined that 558 feet of the mass lay above ground while 1,181 were buried beneath the Carstenszweide. The ice, in effect, had polished the Ertsberg and done a lot of our mining for us. While exposed ore masses due to glacial erosion are not uncommon, the Ertsberg is perhaps the largest such mass ever found.

For three hours, Del and I cracked rocks and examined ore specimens in the talus or fallen chunks at the base of the Ertsberg. One huge angular slab, which had fallen off the main mass since the last Ice Age, weighed 50,000 tons or more.

Though the glacial flow had exposed most of the upper section of the Ertsberg, the northeast flank remained attached to an uneroded portion of the igneous mass from which it was derived. This afforded convenient access to the top of the deposit. We followed a rough trail made by

Gus and Del that spiraled up the attached flank. Other than some smooth, slippery areas of intense glacial polishing, walking on the ore was relatively easy because weathering had removed from the surface the soluble copper minerals and left corrugated ridges of nonsoluble magnetite. Deep grooves pointing south clearly denoted the passage of countless tons of ice. In a few places the glaciers had scooped out small depressions, which enabled small bits of moss to gain a foothold. Otherwise the top of the Ertsberg was completely exposed.

As we approached Gus and three Sansaporese, who were chipping off samples near the top, the visibility deteriorated. Because there were many places high on the face of the deposit where one misstep could mean a 500-foot fall, we returned to camp.

At six o'clock in the evening, after two days of calling, we at last established contact with Kokonao. Much to our relief we learned that John Bowenkamp had arrived on the 16th and had been transferred by charter flight to Biak where he would receive medical attention in the Dutch naval hospital. We were also able to send out a message to Bob Hills, my boss in New York, and to our Dutch partners in The Hague giving them our preliminary estimate of the size and quality of the ore deposit. To avoid premature disclosures, we sent the message in a pre-arranged code:

Freesulph, New York
Hills sixteen Wilson joined overland Carstensz tremendous trailbreak Wintraecken Flint thirteen acres rock above ground additional 14 acres each 100 meter

depth sampling progressing color appears dark access egress formidable all hands well advise Sextant regards.

Wilson

"Thirteen acres" meant 13 million tons of ore estimated to lie above ground, "14 acres" meant the number of tons (in millions) below ground to a 100-meter depth, and "color appears dark" meant that the grade of ore was good. Sextant was the code name of the East Borneo Company.

Saturday, June 18: Gus, the first one out of the tent, reported that the new thermometer on the east flagpole registered 34 degrees. On his way out he cracked ice in the folds of the silk parachute cloth forming the canopy over our wash-up area.

This morning we started a daily routine which would continue for the next two weeks: Climb out of warm sleeping bags and into wet, cold clothing hanging under the canopy in front of our tent; brush the teeth and splash cold water on the face; solemnly down two aralen pills for malaria prophylaxis and one antibiotic for our numerous festering sores; change dressings on infected parts; drink a cup of hot coffee and swallow one general-purpose vitamin pill and two iron pills for altitude, as recommended by Colijn in 1936. After coffee and medication, Gus and Del would head for the radio shack for the early morning conference with Kokonao while Wilson would light a Super Bluet to keep Mardjuki's oatmeal breakfast hot.*

---

* The Super Bluet is a remarkable development of French Alpinists. It is a small bomb of butane, weighing less than a kilo, and when attached to its special stove assembly will burn with a hot flame for two hours. A large

After the Kokonao call, Gus and Del with four of the Sansaporese, known as Bernhard, Jocias, Marinus, and Barnabus, would continue sampling the Ertsberg's summit. Wilson with Gatt, a Sansapor man who wore a ring of metallic tin in his right ear, would sample the large tonnage of ore that had accumulated around its base. Although Gatt and I had no common idiom, it took me no more than 15 minutes to instruct him in his duties as a surveyor's helper and the procedure to be followed in stretching a 50-meter steel tape between survey points.

By one in the afternoon on most days, mist and light rain would make further outdoor work uncomfortable. After changing into dry clothes and a lunch of soup and rice, I would work on my notes while Gus and Del supervised preparation of our growing store of ore samples. With a hammer and small anvil set up on a piece of burlap, the samples were broken down into pieces less than ⅛ inch in diameter, a task at which the Sansaporese were especially adept. In case of loss, we prepared two samples from each location. Each weighing about five pounds, the samples were packed separately in heavy, carefully labeled canvas bags which in turn were packed into five-gallon tins weighing twenty-five to thirty pounds. Shipments of samples were forwarded to Base Camp by the Sansaporese and transferred to Tsinga by mountain Papuans under the supervision of Bob Croin's police officers.

---

supply of these life-savers had been included in the first parachute drop. In addition to their value in cooking food in Mardjuki's quarters, where wet wood and precious little of that was the alternative, and in keeping food hot after it was delivered to our tent, we found that a five-gallon empty oil tin mounted over a Super Bluet made an efficient space heater in our tent during the evening.

Following dinner, we would light the Super Bluet, break out our supply of Remy Martin and Vat 69, and deal the cards for the evening gin-rummy game, which went on for an hour or more. When we left the Carstenszweide after hundreds of hands, the score was a triple dead heat, clear proof of the inescapable laws of chance.

The view during my first morning at the Ertsberg was so spectacular that I decided to interrupt surveying for a few pictures. It was a good thing that I did, for this would be one of only two days during my entire stay when the sun shone past ten in the morning. Gatt and I walked across the damp ground at the base of the Ertsberg, where most of our supplies had been dropped, and climbed to the edge of a level plain stretching to the northwest which we called the upper Carstenszweide.

From a position about one-half mile northwest of the Ertsberg, I looked back to the southeast. The rays of the sun, which hit the top of the Ertsberg at 0945, cast a shadow close to the edge of the cliff I had climbed two days before. I could see Gus Wintraecken and two of his Sansaporese in lemon-yellow rain jackets climbing near the top of the ore deposit. They looked like yellow ants climbing over a dark lump of sugar. In the distance was snow that had fallen the night before on the lower slopes of the Carstensz peaks.

While I was burning up one of my few remaining rolls of film, clouds began rolling up the valley from the south. Shortly after 1000, my view was abruptly cut off.

Jean Jacques Dozy (left), Franz Wissel, and A. J. Colijn rest on the glacier near the summit of the 16,500-foot Carstensz Top, object of their 1936 expedition to the New Guinea highlands. Geologist Dozy's report on his discovery and sampling of the nearby Ertsberg outcrop inspired Forbes Wilson to undertake his 1960 expedition.

Moses Kelangin (in T-shirt), a mountain tribesman and Wilson's invaluable guide, is flanked by bearers he recruited from mountain villages. Armed with bows and arrows, the bearers wear only jewelry and *kotekas* fashioned from wild gourds.

The Wilson expedition forges up the Mawati River in 50-foot dugout canoes carved from huge trees. The 14-canoe fleet required 44 paddlers from coastal villages.

From poles laced together with rattan strips, Irianese tribesmen construct a passenger bridge over the treacherous Aghawagong River near Base Camp.

Pedaling furiously, former Dutch marine Jan Ruygrok generates power for the jungle radio while Molle (center) handles transmissions and Takim (right) prepares food.

Del Flint, Wilson's chief aide on the venture, poses next to one of the pygmy Irianese who inhabit the Irian Jaya highlands and who often smoke local cigarettes lighted at both ends.

Base Camp, from which the final assault on the Ertsberg was launched, was established in a small clearing at the junction of the Oetekenogong and Aghawagong rivers a mile above sea level.

Ten miles to the north and 7,000 feet higher, the exploration camp near the base of the Ertsberg was a rugged two-day climb from Base Camp.

Construction of the 63-mile access road from a port near the coast to a canyon below the Ertsberg took two years and was the most arduous part of the $170-million development project. *Left:* An initial 1970 "pioneer" road snakes up the sheer Darnell Ridge, whose grade often exceeds 70 degrees. *Below:* Bulldozers slice millions of tons of earth off ridge tops to reduce the grade for truck traffic. *Above:* The road tunnels 3,627 feet through Mount Hannekam.

◄A jet-black monolith containing 33 million tons of high-grade copper ore, the Ertsberg soars 586 feet through the mists above a 1967 drilling camp. Some 1,181 feet of the deposit lies buried beneath the grassy meadow carved out by an Ice Age glacier.

Sixty miles north of the coast at an elevation of 8,500 feet,
the finished access road traverses a fog-shrouded ridge crest.

Fifteen acres of thick mangrove swamp were cleared by chainsaws for
the port facility on the Tipoeka River.

At the fully operative port facility a barge stands by to receive copper concentrates transported down from the mine by slurry pipeline. These concentrates are shipped to smelters in Japan and West Germany

Near the base of the Ertsberg, in 1972, foundations are under construction for tramways to carry mined ore from the top of a sheer headwall to the mill in a canyon 2,500 feet below. The barracks perched on the edge of the headwall (to the left of the construction site) were affectionately known as the Stone Age Hilton.

Left: A tram car carrying ten tons of ore chunks begins its run down the mountain. *Above:* The car releases its load, which will be crushed at the mill. In concentrated form, the copper ore is mixed with water and fed into the pipeline for the trip to the coast.

The 4½-inch pipeline, longest of its kind ever built, is laid along a shallow trench cut through the coastal jungles.

Eight years of mining has leveled the Ertsberg's towering outcrop, and a deep open-pit mine now eats steadily away at its underground mass. This aerial view, showing the tram machinery in the foreground, was taken from a position not far from that shown on the book jacket. ▶

Left: The town of Tembagapura, built in 1072 by Freeport on a rocky plain at the base of 14,500-foot Mount Zaagkam, serves as home for over 3,000 workers and dependents. *Above:* The workers commute by bus to the mill and mine sites ten miles to the northwest.

To supplement Tembagapura's imported food supply, a Freeport helicopter makes a weekly flight to Beoga, a remote village north of the Carstensz peaks, to pick up locally grown bananas and other produce.

Protected by slickers from the ubiquitous rain and fog, Indonesian miners descend along a road leading to the bottom of the open-pit mine, now over 500 feet deep. The mainly Indonesian work force is recruited from throughout the 3,000-mile archipelago.

A mining crew rides a locomotive along an adit used to extract ore from the recently discovered Ertsberg East deposit. Less than a mile from the original Ertsberg, this underground mass could contain three to five times as much ore.

Despite the chill weather, inhabitants of a small squatter settlement on the outskirts of Tembagapura still wear tribal dress.

Two local villagers, who lived a virtual Stone Age existence prior to Freeport's arrival, have now been trained as mechanic's helpers.

A nurse at the Tembagapura hospital treats a foot injury. The hospital regularly provides care to residents of nearby villages, especially following periodic bow-and-arrow inter-tribal battles.

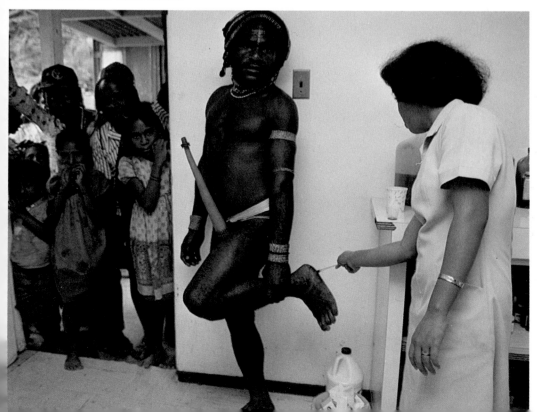

Forbes Wilson (in red hardhat) discusses engineering problems in 1972 with Del Flint (to Wilson's left), Bob Wernet (next to Flint), and officials of the Bechtel Corporation, a large San Francisco firm responsible for the construction of the Ertsberg project.

Wilson poses in 1976 with Freeport geologist Frank Nelson, who supervised exploration of the Ertsberg East deposit.

Not long after returning to our surveying and sampling, I learned, once again, that one can never take the continually surprising New Guinea environment for granted. Standing on the mossy ground at the base of the Ertsberg, I had completed cutting an ore sample weighing about 15 pounds, which I put in a canvas bag. I casually gave it a toss in the direction of a pile of other bags about 20 feet away. The throw was miscalculated by about three feet. With an audible "plop," the bag disappeared into the moss, leaving a neat round hole that immediately filled with water. After sloshing around in muck up to my waist, I found the bag two feet under the slimy surface. It was the same invisible swamp into which the drum of sugar had plummeted during the June 6th air drop.

Over the next 10 days, Gatt and I completed cutting 38 samples from the talus and making a survey of the Ertsberg's formidable dimensions. One day, while peering through my survey instrument, I noticed a prominent splash of green in the background. Lowering the instrument to get a better view, I was surprised to find myself looking at a high limestone cliff about a half mile to the northeast. The green exposure, possibly 10 feet wide and running down the cliff about 100 feet, was obviously a stain of malachite, a carbonate of copper. Gus later discovered a second larger malachite stain on another cliff two miles from our camp. The cliffs between us and the stains were too steep to climb and we never were able to conduct a detailed investigation. Not for a decade would we learn that the malachite stains were surface expressions of what would be called Ertsberg East, an underground copper deposit as rich as the Ertsberg and several times as large.

Saturday, June 25: Today a full camp holiday was declared. The long shuttle service between the Ertsberg and Base Camp had taken its toll on the hard-working Sansaporese. Nearly all had developed infirmities, especially swollen glands in the groin. After trying several medications, Gus and Palijama, our male nurse, discovered that most of the ills responded favorably to penicillin within 24 hours. While the Sansaporese recuperated in their tents, Gus, Del, and I took off at dawn to explore the upper Carstenszweide. In an hour, we reached its southern edge, where I had taken my pictures the previous week. The plain of thick yellow grass stretching in front of us looked much like a ripening wheat field.

We expected to find travel through the yellow grass easy. After less than 100 feet, though, we were jumping from one grass hummock to another to avoid floundering hip deep in soft muck. Forward progress in this morass was like breaking trail in deep snow.

The plain was crisscrossed with small game trails, probably formed by the night foraging cuscus avidly hunted at lower elevations by mountain Irianese. At one point Del Flint came upon a small covey of quail-like birds which my friend Tom Gilliard subsequently identified as a species of Asian quail. They were so uninhibited by the presence of humans that they could easily have been captured alive.

We soon found more solid footing was available along the west side of the upper Carstenszweide where a few tufts of grass struggled for existence between small limestone fragments. We crossed a small stream flowing down from an even higher glacial moraine which Colijn had named Grassweide.

A short distance beyond the stream we came upon what Colijn had called his Alpine Camp. Even after 24 years it was easily identified. The ground was littered with sardine tins and other metallic containers of food. We were surprised they had not rusted away in this wet climate. The reason, of course, is that oxygen is scarce above 12,000 feet elevation and metallic decay in such an environment proceeds at a more leisurely pace than at sea level. At Biak, for example, the more recent relics from World War II are rapidly disintegrating on the beaches.

We could appreciate the name Colijn had given to this spot. Surrounding us on three sides were towering mountains of white limestone. We could now see across the upper Carstenszweide to the northeast the entrance to the valley that Dozy and Colijn had crossed before ascending the north wall of the Carstensz Top.

As we approached, the head of the valley was lit by sunlight slanting over the Carstensz glacier. We came upon bizarre vegetation: huge stalks a foot or more in diameter and 12 feet high with fernlike leaves. At first they seemed a variant of tree ferns we had seen frequently at lower elevations, but unlike tree ferns, they did not have hard woody trunks. When I hit one with my geology pick, the shaft sank in up to the hilt in soft green pulp like that of some vegetable. I struck another that appeared dead or dying. Blood-red liquid flowed out of the wound. It was a disturbing sensation. I felt as if I had wounded an animal.

The plants appeared to be members of some ancient order. Indeed, they looked like flora portrayed in textbooks illustrating conditions millions of years ago when coal deposits were being formed. When we sat down to rest in the midst of a clump of the plants, the scene seemed so otherwordly I found my thoughts dwelling on the possibility that prehistoric animals might still be living in this isolated valley. (Others have had similar thoughts about such places. See Appendix Note 4.)

The narrow neck of the valley leading to the Carstensz Top was now only a half mile away and I forgot the mysterious plants while we spent a futile hour struggling across the bog. We succeeded only in floundering less than halfway to our objective in black ooze. There was no time to try a detour. Exhausted and covered with slime, we headed back down the valley.

During the return trip, we had a high-level view of the daily contest between moisture-laden clouds sweeping in from the hot, damp south coast and the cold winds blowing in reverse direction from the Carstensz ice field. Long before we reached the south end of the upper Carstensz-weide the clouds won the battle. Swirling fog blotted out the scenery and the usual afternoon misty rain was falling when we reached our camp.

As we neared the end of our engineering studies, we gave increasing thought during evening strategy sessions to the task of getting back to the coast. It was obviously more than merely climbing down what we had climbed up.

Gus, Del, and I probably could have struggled out by ourselves with enough food in our knapsacks and from the supply depots to survive. However, we had accumulated over 600 pounds of ore samples, and if they did not get to the coast also, the expedition would be a failure and all of our efforts would be in vain.

We felt secure that the Sansaporese would be able to finish transporting the samples to Base Camp and that enough packers could be recruited to move them to Tsinga. The problem was Tsinga to Praubivak, the northern extreme of canoe travel on the Mawati River, where the samples would be transferred to canoes manned by coastal Irianese. We had already utilized as bearers nearly all of the available adult male population in the vicinity of the Tsinga-Praubivak route. Each had received an ax blade a parang, and an assortment of other goods. Having no sense of Western-style wealth accumulation, the Irianese would likely see no need for more than one of an item like an ax blade. What could we use as additional inducements?

The answer, we decided, was khaki shorts and white T-shirts. Father Koot had advanced the idea of using these garments as inducement goods earlier in the trip. Since white men and such a respected indigenous leader as Moses wore them, Father Koot argued, shorts could well replace the gourd as the preferred apparel of status-conscious mountain Irianese. While intrigued with the notion, we had initially opted for more conventional compensation. The time had clearly come, though, to experiment with more novel items.

A preliminary sounding of opinion among the mountain people recruited to move initial shipments of ore samples from Base Camp to Tsinga indicated the experiment would be successful. Our trip back, we figured, would require no less than 150 pairs of khaki shorts and T-shirts, which would have to be air-dropped at Tsinga. Since we expected to be coming through Tsinga in early July, the air drop would have to be completed by late June.

During one of our daily talks with Kokonao, we asked that instructions to this effect be relayed to Kroonduif Airlines at Biak. Kroonduif's first reaction was that it would be impossible to make the drop before the middle of July. We responded with a sharply worded cable stressing the gravity of our situation. Two days later we were told that Kroonduif would complete the drop on June 24th. On the evening of the 24th, we were told the drop had been made on schedule. My field notes at the time indicated my confidence that our trip to the coast would be uneventful. This prediction proved to be so wide of the mark that in retrospect it can only be characterized as complacency run amok.

Monday, June 27: This was to be our last full day at the Ertsberg. All of us were anxious to leave. Living for several weeks in constant dampness and working in wet clothing had produced a rheumatic stiffness in our joints. The bland, monotonous diet of canned food, mainly soups, had caused us to lose weight, leaving us looking like gaunt old men. And while there had been no open animosity, the close quarters and unchanging routine were producing a general feeling of irritation. We were just getting on each other's nerves.

We sent the last of our samples down the mountain toward Base Camp and sorted out our supplies and equipment to see what should be taken and what should be left behind. The unused Super Bluets, lethal if accidentally dropped in village campfires, were buried in a pit dug in the wet moss along with other goods that, if discovered by the Irianese, could detract from the attractiveness of our regular inducement goods. We buried separately our remaining liquor supply—two bottles of Remy Martin—in a place where we thought they could be easily recovered by members of a future expedition. (Regrettably, they were buried in a landslide several years later.) In the evening, we had messages relayed to New York and The Hague that we were about to begin our return to the coast.

Chapter Seven

Tuesday, June 28: The day began with what turned out to be deceptive propitiousness. Del and I with four Sansaporese left the Ertsberg camp at 0745 in bright sunlight. For the first hour and a half, we picked our way down the steep headwall to the Aghawagong. During four weeks of shuttle service between the Ertsberg and Base Camp, the remarkable Sansaporese had made vast improvements in the trail by erecting in the steeper places long strings of rattan-tied ladders, some more than 150 feet in length. We made it to the bottom with ease. But then our troubles began. At the foot of the cliff, I slipped on a rock and tumbled into an icy pool almost over my head. I must have sounded like an enraged sea lion as I surfaced on the opposite bank, cold, very wet, and in no mood to join Del Flint in laughter. Just a short time later, Del, who was carrying a heavy packsack, lost his footing crossing the river and was swept over a 10-foot waterfall, his head barely missing a large rock in the drop. He fell into a deep pool and was pressed by the swift current against a ledge. His knapsack was pulling him down, and it didn't look as if he would be able to get out. In near panic, I rushed across the stream and helped him climb up the bank.

Just a short way from Camp 12, I suddenly became nauseated. It was the first time I had felt really ill since leaving the coast. My knee joints were very stiff. After an especially painful half hour scrambling over large boulders on the streambed, I reached the camp, stripped off my wet clothing, and wrapped up in two heavy blankets. We thought the Sansaporese with our bedding were directly behind us. Anticipating their imminent arrival, we lay down on the round poles that served as the base of the sleeping platform.

Three hours later, the Sansaporese were nowhere in sight. Searching for something to cover the unyielding poles, we found two bags of mouldy wood shavings. We were in the process of spreading this damp, evil-smelling stuff when the packers at last staggered into camp. They were again showing the strain of continuous hiking between Base Camp and the Carstenszweide. The undistinguished supper of cold corned beef and tea did nothing to mitigate my concern about the inauspicious start of our return journey.

Wednesday, June 29: The morning was clear and we were on the trail by 0725. For the first two hours, I felt fine, but by midmorning I began feeling very tired and nauseous. My knee joints creaked like rusty hinges and the muscles in my upper thighs cramped painfully. My spirits were only slightly revived upon my arrival in Base Camp when I was greeted by Moses, who had long since recovered from his illness.

As I was stripping off wet clothing, I heard the now familiar sound which mountain tribesmen make in moments of excitement. The sound, which is quite effective in gaining one's attention, is really a mixture of such elements as the baying of fox hounds and the crying of sea gulls. To the east across the river I saw a group of 25 Irianese strung out in single file and moving along the far bank in their characteristic dog trot, leaning forward with knees slightly bent and gracefully traversing uneven ground with the greatest of ease. They were from Jonkogomo, an advance section of the group who would accompany us out.

We had to wait again for the arrival of our baggage, but here at Base Camp we could lie down on camp cots. For the first time in weeks, the temperature was well above freezing. Bob Croin's cooks gave us a rib-filling three-course meal including thick Dutch soup, corned beef and rice, and fruit cocktail for dessert.

Thursday, June 30: Today was a day of rest and I needed it badly. I ached all over. While a laundry detail went to work on our mud- and sweat-soaked clothing, we were able to have our first adequate bath in weeks in the waters of the Oetekenogong.

A party of 10 packers left at 0800 for Tsinga with five five-gallon tins containing ore samples. Three quarters of the product for which our expedition had spent $120,000 was now ahead of us. At 1100 another small party came into camp from Amkaiagema. While this increased our bearer support to 35, Moses and Bob Croin felt we could not proceed without additional manpower.

An equally serious problem was the surplus supplies, mainly foodstuff, remaining at Base Camp. If we simply abandoned the supplies, they might cause intertribal warfare among the small groups living in the hills near Wa. We decided to move the material to Amkaiagema, where they could be added to Father Koot's missionary stores. Moses set out to recruit a group of especially trustworthy and dependable men to serve as bearers.

In the early afternoon, Gus arrived from Camp 12. He reported that the Sansaporese in his group had badly infected feet and would not be able to travel on the following day. When they limped into camp several hours later, it was clear he was right. We decided to lay over for another day before attempting the three-day walk to Tsinga.

Friday, July 1: We were up at 0630, but there was very little to do except wait. No new bearers arrived. We couldn't force the Sansaporese, who had been so loyal and important to our mission, to travel with infected feet, but neither could we sit here indefinitely. After innumerable tiresome games of gin rummy, we bathed in the river and thawed out our rheumatic joints in the sun. We were all more worried than we cared to admit.

We turned on the radio to see if we could pick up any news. We learned that Molle and Ruygrok had left Tsinga on June 30th to arrange for canoes to meet us at Praubivak. We were also told that there would be a police boat at Omowka on July 16th to pick up Bob Croin and the police officers and that the coastal steamer, Oranje, would call at Agats on July 21st to pick up Gus, Ruygrok, the Sansaporese, and all those returning to Sorong.

Late in the afternoon, additional bearers began drifting

into camp, and by sunset we had enough to resume travel. Among those arriving was Martinius, my porter on the inward trip from Tsinga, who indicated that he wanted to pack my things all the way back to Praubivak.

Saturday, July 2: Del and I were off at 0730 in the van of 52 bearers, 19 Sansaporese, and a mixed group of 12 Eurasian and Indonesian camp foremen, cooks, and attendants. Gus, Moses, and Bob Croin and his police were getting ready to leave shortly after us. The canvas top of the godown had been removed and the surplus supplies had been formed into neat piles for Moses' special bearers, whose arrival was expected momentarily. At last, everything seemed under control. Not until late in the day did we learn of the looting and near riot that occurred soon after our departure.

Del and I crossed the river and headed upstream toward the Zaagkam and Camp 10. In a brisk climb of an hour and a half we reached the level spot at the base of the Zaagkam that would later become the site for Tembagapura, the town where the work force would live during the Ertsberg's development. While we were resting here, Moses' 12 special bearers trotted by in single file.

We attacked the steep ridge leading off the southwest flank of the mountain and after four hours of steady climbing reached the top of the ridge, 4,600 feet above Base Camp. The travel had not been too difficult but it was quite apparent that two days of rest had done us a lot of good. While I could not climb for long periods without excessive puffing, I revived promptly after each rest period, which I had been unable to do on the first two days coming down from the Ertsberg.

Groups of packers passed us as we rested in the wet moss on the crest of the ridge and snacked on chocolate bars and rectangular lumps of food concentrate, which Gus referred to as grease cookies. When I returned to America, Dr. Watson reported an increase in my cholesterol level despite a substantial loss in weight. The grease-cookie diet was the probable cause.

At about two o'clock in the afternoon, Del and I arrived at Camp 10 some 900 feet below the top of the ridge. It was raining and, as our camp equipment had not arrived, we moved into a small, hastily constructed lean-to where 20 or more mountain people were crowded around a smoky fire. For two hours we tried to keep warm and dry without becoming cooked and smoked, a process that involved a nearly continuous movement of personnel from the outer fringe of the lean-to toward the fire and then a gradual retreat to the rear again.

Gus arrived with some of our camp equipment at five o'clock in the afternoon. In the strong, wind-driven rain, it took us some time to spread the main tarpaulin of our tent. After we had gained some protection from the weather, Gus told us of the wild scene at Base Camp after we left.

If Moses' special bearers had arrived on time and removed the surplus goods, nothing probably would have happened. But they were late. As the mountain tribesmen watched the main body of the expedition depart and the few remaining members getting ready to leave also, some apparently thought the piles of surplus materials were being left behind. Sporadic pilferage of canned goods quickly turned into mass looting as dozens of Irianese stuffed as much as they could in their net bags. The melee did not subside until Bob Croin and his one remaining police officer fired their guns into the air. The Irianese, though, refused to give up their booty or pick up the loads they had previously agreed to carry to Tsinga. After some negotiation, Moses reached a settlement that allowed the looters to save face. They would perform their bearer obligations, but no further demands would be made for the return of the stolen goods.

Several hours were required to clean up the mess and reorganize the bearers with their assigned loads. Most were so heavily burdened with a miscellaneous assortment of purloined canned goods, parachute cloth, and discarded burlap sacks that they did not begin to arrive in Camp 10 until late in the afternoon. The last to arrive, long after sunset, was Martinius, loaded down with my bedding, a silk parachute, a wool blanket, and many tins containing foods and condiments with which he was completely unfamiliar. He was exhausted and contrite and through sign language attempted to convey his shame and apologies. I doubt he received any comfort from my churlish response. Moses later severely berated him and discharged him from the expedition.

I spent the evening shivering near the weak fire we had managed to bring to life in one corner of our tent. It rained all night and those of us who slept on the windward side of the tent woke in the morning under a cover of blankets made sodden from the wind-driven spray.

Sunday, July 3: Today we were scheduled to negotiate in reverse direction the section of the trail that on the inward trip had given Del Flint so much trouble. I knew that to Del the prospect of confronting again those 2,000 feet of precipitously cascading waterfalls below the Zaagkam had become, during the past few weeks, a serious physical and psychological barrier. So traumatic had been his first experience that he was not certain he would be able to make it through a second time.

We left camp at 0740 well armed with stout walking sticks. A thirty-minute walk brought us to the top of the waterfalls. We inched our way down, bracing our walking sticks in the cracks and grooves to permit us to locate footholes in rock fractures. I could see the growing apprehension in Del's face. When we reached the steepest section of the incline, however, we found that the most troublesome cliffs had been equipped with crude ladders by the policemen during their shuttle service over the past month between Base Camp and Tsinga. Del's fears vanished.

My own fear about this part of the trail had been broken a month earlier. Despite my concern about Del and the Base Camp incident I had felt quite jaunty since the beginning of the day. Unfortunately, I was so relaxed that halfway down I slipped on a slope of slimy red shale and slid along a smooth chute toward a waterfall that dropped off a cliff at least a hundred feet high. The 20 mountain people watching the scene cried out in anguish. But even though I seemed to be sliding toward eternity, I felt curiously calm. At the top of the waterfall, I could see a flat protective shelf about six feet wide on which, based on a quick calculation of the friction factor of wet britches on wet rock, I felt certain I would come to rest. The calculation proved accurate. When I stopped, I peered over the cliff where my companions had thought I was destined to be dashed to bits. They threw me a rope, and I scrambled back to the narrow ridge of rock at the edge of the jungle.

At the foot of the waterfalls, the trail continued mainly along the streambed, but in avoiding sudden drops it wandered off occasionally into dense jungle growth. Despite my nonchalance, my fall had left me weak and shaken. Like my terrifying bridge crossing on the way in, it had put more of a strain on my system than I had first realized. Walking through the jungle, I fell repeatedly in the twisted vines and heavy vegetation. Del, who was walking behind me, mentioned after one of the falls that a cut near my ear was bleeding. I couldn't feel any pain, but on touching my neck I found it was covered with blood.

When we reached the Nosolonogong River and the site of Camp 9 at about 1400, I was covered with mud and slime and still bleeding from my wound. I walked into the river and sat down in water waist deep to remove some of the grime. Del and Paliama, our male nurse, came over to look at my head. Paliama made some comment in Malay and opened up his medical kit. "You have a small hole in your head behind your left ear," Del said. Apparently, in one of my falls, a sharp projection, perhaps a pointed stick, had punctured my head in one of the few points of the body where there are no nerves. While I still felt no discomfort, there was a definite hole in my head, something that is not advisable to have on a long trip through the jungle. It continued to bleed in a slow, oozing fashion.

Paliama cut away some hair, dusted the wound with a sulfa drug, and applied a Band-Aid.

The weather remained clear and after a brief rest at the river we decided to push on and make up previously lost time. We climbed 1,000 feet above Camp 9 to the east end of the vast landslide encountered on the first day out of Tsinga. Before we passed the midpoint of the landslide, it began to rain, which made going across the rubble of rocks slippery and dangerous. We were forced to spend a wet and uncomfortable night in a makeshift camp, but at least we had reduced the length of the last day's travel to Tsinga.

Monday, July 4: While the other members of our party broke camp, I took steps I hoped would facilitate travel through the jungle. I cut off my heavy khaki pants at a point just above the knee, thereby converting pants to shorts, which would allow my knee joints to function more efficiently. I strapped on a set of canvas puttees to protect my shins from scratches and bruises, and, digging into my packsack, I found my ice creepers or crampons and strapped them firmly to my boots. The creepers, which had been among the items in my duffel bag, had metal prongs normally used for walking on ice. I hoped they would facilitate my passage along the soft logs that spanned many of the streams along the trail.

When I started up the trail, I felt like a new man. The crampons allowed me to walk freely and worked perfectly on the logs. We climbed the ridge northwest of Tsinga where I had rested a month earlier after negotiating the knife-edge ridge of sandstone.

After a brief rest and a few small chocolate bars, we

started down. Descending the ridge was more difficult than the original ascent. If I went down backward, as one would normally go down a ladder, I found that I could not pre-select footholds and had a feeling that if I ever started sliding I would be cut in two by the sharp ridge.

At the foot of the cliff we came upon a bog which I remembered from the inward trip because of the unusual foliage. Continuous foot traffic through this area for the past month had churned the trail into a knee-deep morass of mud, the worst we had encountered to date. At the first stream where I could wash off the accumulated clods, I also removed both puttees and crampons. In mud and loose rocks, they were definite liabilities.

Shortly before noon we arrived at the Magogong crossing, 1,200 feet below our next camp. The trail uphill was not hard but we were tired. After two hours of climbing we walked through the village of Jonkogomo. We could see the Dutch flag flying from the top of a dead tree marking nearby Tsinga camp. Advance word of our arrival had reached the camp. As Del Flint and I emerged through the brush on the outskirts of camp, Bob Croin's three police officers moved into position with Boer War muskets and cut loose with a triple salvo of welcome. The surprisingly loud shots were followed by excited cries from 40 or 50 mountain villagers watching the performance. It was a fitting end to a rather fatiguing Fourth of July.

The police officers we had left in Tsinga early in May greeted us like conquering heroes. They pressed upon us bundles of mail that had come down with the air drop of goods on June 24th. We were fully occupied until dusk reading letters that had been sent from Europe and America weeks before.

In the evening, after our first good meal since leaving Base Camp three days before, a group of villagers started singing around the camp fire. I regretted not having a tape recorder to catch the pleasant harmonies of their simple songs. Some of the savage tribes of Danis in the Baliem Valley well to the north of Tsinga are reported to accompany their native music with the gnashing of teeth but the Irianese here were far more melodious.

Tuesday, July 5: This was the day we had announced that the bearers who had brought us through from Base Camp would be paid off with our standard inducement goods of axes and parangs and that those who agreed to accompany us to Praubivak would be rewarded with the coveted khaki shorts and T-shirts. The event generated a great deal of local interest. By 0900, our camp was thronged with 100 to 150 villagers. The crowd included not only bearers, and in many cases their wives and children, but many others who simply wanted to watch.

Busy with other matters, I did not notice until late in the morning that the camp was deserted. All the villagers had disappeared. I soon found out what had happened. After the bearers from Base Camp had been paid, the prospective bearers to Praubivak had moved forward to receive the khaki shorts and T-shirts. But there were no khaki shorts or T-shirts to give them. The 60 bundles of goods air-dropped on June 24th, Moses discovered when he began opening them, contained only rice and a miscellaneous assortment of cheap knives and trinkets. Rice was already a surplus and valueless commodity in Tsinga because, lacking any cooking or eating utensils, the villagers never ate it. The air drop had just added another ton to our two-ton supply. The police officers in Tsinga in fact had been sleeping on rice bags since early May.

Moses tried to pass out the knives and other trinkets as a substitute, but the villagers quickly recognized that these goods were even inferior to earlier inducement goods. As soon as it became apparent that the white men had not lived up to their word, the villagers took off into the hills.

We later confronted the head of Kroonduif Airlines with what had been a deliberate disregard of our quite specific air drop instructions. His explanation was that he assumed that after a month in the mountains we must have been a little out of our minds. He could not imagine how we could actually need 150 pairs of khaki shorts and T-shirts. But he figured we probably did need food and a fresh supply of inducement goods.

Our expedition again seemed on the verge of collapse. For a time, we couldn't think of anything to do but stare into the mist and curse the fools at Biak who had left us in the lurch. Once again, however, Moses came to our rescue. He knew we had been betrayed by our suppliers on the outside and that we had intended to keep our word. As soon as our bearers left, he went off in search of replacements. By late in the evening, he rounded up 20 men who, for the usual inducement goods, would take us to Amkaia-gema. But we still needed 50 to 60 more to move the ore samples, as well as food and baggage, to Praubivak.

Wednesday, July 6: During the night, Moses left on another recruitment trip. Late in the afternoon, after we had spent many anxious hours waiting, he returned with about 25 local boys who seemed no more than 12 or 13 years old. More boys were to arrive the next day. Since most of their fathers had already served as porters and had obtained as much of our inducement goods supply as they wanted, Moses had turned to the next generation. The boys would go with us to Amkaiagema, where Moses would try to sign on a replacement contingent of young bearers.

Thursday, July 7: Gus, Del, and I met with Moses to analyze what we had to move and who we had to move it. It was clear we would have to reduce our cargo to the absolute minimum. We had to take our 20 five-gallon tins of ore samples, each of which weighed 15 kilos or 33 pounds and constituted one load. After a rigorous separation of the essential from the merely desirable, we cut our food and medical supplies to 15 tins, and our bedding, eating equipment, and miscellany to five loads. Instead of 60 packers, our requirement was down to 40.

As priorities were established, Moses assigned the loads to his squad of young packers. Each boy took possession of his individual package or five-gallon tin, manufactured a rattan harness to fit his head and shoulders, and gave every evidence of complete acceptance of responsibility.

Friday, July 8: We broke camp early in the morning. All of us knew that it would be a long and difficult day. No one was there to see us off, but we were comforted by the fact that we were again moving forward and also by the fact that most of the journey to Amkaiagema would be downhill.

We were a ragged-looking group. The young packers

started off bravely but they were straining under their loads and needed frequent rest stops. The local police officers had agreed to carry their own personal effects, but it was quite obvious they were not happy about this loss of face. After nearly six weeks of idleness in Tsinga, they were not in good trail condition. The Sansaporese were still limping on infected feet—an army of walking wounded. The expatriates, Wintraecken, Flint, and Wilson, had continued to gain strength since leaving the Ertsberg, but as bearers of anything but our bodies we were a total loss and therefore unable to make any effective contribution to the forward march.

Like Del Flint and his return to the waterfalls five days ago, I would confront today my own physical and psychological obstacle: that fearsome bridge over the Nosolonogong River I was just barely able to cross on June 9th. I approached the ordeal, though, with confidence. First, after learning of my problems, Moses had issued orders to reconstruct the bridge with new and stronger supporting timbers and guidelines. Second, I had my ice crampons to provide a more secure foothold on the logs. I intended to march across the new bridge in upright fashion and prove that an expatriate, properly shod, could do just as well as a mountain Irianese.

As we came down the final slope to the Nosolonogong, now roaring with twice the water volume of early June, I put on my crampons. Though ineffective on rocks, I wanted to get the feel of them again. When I saw the bridge, I felt my first twinge of concern. There was no visible evidence of reconstruction. As I came closer it was obvious that not only had the timbers not been changed but, if anything, the bridge had deteriorated since the inward trip. I subsequently learned that Moses' instructions had not been clearly understood and that a perfectly good bridge somewhere on the east side of the Tsing River had been torn down and rebuilt.

With dismay I watched as 15 or 20 of our small packers walked over the bridge, not even bothering to touch the drooping lines of rattan on the sides. Del Flint, the next to proceed, said he would try to make an upright crossing and started off bravely. About halfway across one rattan line parted. Del tilted dangerously to the left, regained his poise, and somehow continued to the far side. I was now faced with an inferior bridge with a finger guide of rattan available only on the right-hand side. After some mental conditioning, I started across. I had moved no more than 10 feet when the right-hand rattan strand parted. I was left without any physical or psychological support. The crampons, meanwhile, were turning out to be more hindrance than help. They were effective on ice and soft wood because they bit into the surface. The logs on this bridge, though, were an extremely hard wood called ironwood that is so heavy it does not even float. The crampons couldn't penetrate the logs. They made my footing even more uncertain than boots.

Using every muscle in my body and a lot of native instinct I didn't know I had, I slowly backed up and returned to solid ground. While the bridge seemed even less able than before to bear up under my weight, I saw no alternative but to cross in my old straddle and leapfrog fashion. I started out again. Each time I leaped forward the structure shuddered and swayed alarmingly. Somehow I made it to the other side, emotionally and physically drained. I took off the crampons and hurled them into the river, which somewhat relieved my tension.

We arrived in Amkaiagema late in the afternoon. We were greeted by Takim, our cook, who had stayed in the village after recovering from malaria. Our packers were paid off and disappeared northward. Only two days travel from the head of the Mawati River and the canoes of the coastal natives, we were again left without anyone to carry our loads.

Saturday and Sunday, July 9 and 10: Moses went into the hills on yet another recruitment drive. Soon new Papuan boys began walking into camp. On Sunday afternoon, as the new retinue of packers was being instructed, Del, Takim, and I decided to push on to Kelangin. We continued the next day to Camp 4 where late in the day Moses, Gus, and the other packers caught up with us.

Tuesday, July 12: Our spirits were high as we started out on the trail today. Our troubles at last seemed behind us. The soft and flabby creatures of early May and June were now lean and tough, and we glided with ease over ground that had seemed almost impassable on the inward march. We expected to spend the night on a sand spit on the upper reaches of the Mawati where Jan Ruygrok had erected a camp to accommodate the canoe men from Omowka who were paddling north to meet us. When I spotted the Mawati, I deviated from the trail, went down a steep cliff, and dived into the river. For a mile or so, I floated down with the current, finally grounding on a gravel bar in midriver. When I stood up, I realized I had just passed through the area where we were afflicted with leeches on the inward trip. I looked at my arms and legs, but this time none were in evidence.

When Del and I reached Belakmakema, we didn't recognize the area. It was the peak of the summer monsoon and had been raining continuously for two weeks. The river where we had dragged small canoes through rocks and shallow rapids was now a torrent. The packers who had preceded us were sitting on the north bank. We soon learned that although it was only a mile to our expected rendezvous with the canoes, they refused to cross the raging river to pick up the continuation of the trail on the other side.

Del and I decided that the best way to break the impasse was to make the crossing first. The going was not too bad by our recently acquired standards, and by the time we were on the far shore we could see a thin line of people entering the river. On this last leg of our journey my seventh and last pair of boots finally disintegrated. I limped into our final river camp wearing a pair of Japanese sandals which still survives.

Del and I were swimming in a deep pool near the camp as our baggage and supplies arrived from the north. By some miracle of timing their appearance coincided with the arrival of canoes from Omowka. The packages were transferred directly from the backs of the mountain Papuans to the big cargo canoes. Having experienced so many frustrating delays and mishaps at Tsinga and Amkaiagema, this rapid transfer seemed ordained by divine guidance.

Wednesday, July 13: The canoes had been loaded and

were ready for an early start. It had taken us two and one-half days to reach this point from Omowka on the upriver trip. Today, with the river in full flood, we expected to return in just five hours.

Our fleet consisted of 21 large canoes, each with four or five paddlers, and two small high-bowed ocean-going canoes, which accommodated Bob Croin and the police contingent. In the swift current, the only paddlers with anything to do were the men in the stern, who acted as helmsmen.

About an hour out from camp a canoe with an elderly man in the bow was maneuvered into the lead position. The old man stood up and in a high-pitched voice started the refrain of a song. At a signal all 93 river men joined in the chorus which ended in a perfectly timed shout. The shout was quite startling to us and to nearby birds, which swarmed out of the trees. The songs continued for an hour or more.

When we reached a point about one-half mile upstream from Omowka, the canoes stopped and the old man who had led the singing made a brief speech. The canoes lined up in three groups, five abreast. Moving to the head of the fleet, the policemen immediately fired two quick salvos from their Boer War muskets. At the signal every paddle hit the water. Digging furiously the fleet raced past the town as the entire population cheered us on from the high bank. We were moving so fast that the canoes could not be stopped and turned back until we were a quarter of a mile downstream from the village landing.

In the celebration accompanying our return to Omowka,

I did not think until later to attend to the serious condition of my left leg. Since the 3rd of July I had been traveling in a pair of simulated shorts. The canvas puttees I had first put on for lower leg protection soon became waterlogged and were discarded. This was probably a mistake. Somewhere below Tsinga I remember passing through a patch of thorns or nettles and shortly thereafter my left leg from ankle to thigh became covered with red spots. They began to fester and did not respond to penicillin or sulfa salve. For the trip downriver, I had covered my leg with a large bandage made from a bed sheet. When I arrived at Omowka and removed the bandage, I saw that some of my wounds were beginning to resemble large tropical ulcers. In desperation, I liberally applied Desenex—the specific for athlete's foot. The next morning my wounds were healing nicely, proving that my problem, as is so often the case in moist tropical climates, was parasites rather than infectious germs.

Thursday, July 14: In our radio chat with Kokonao this morning we learned that the DeHaviland Beaver would land in the Koperapoka, below Omowka, at 1000 on the following day to pick up Del and myself and our set of original samples. We also learned that Gus, Ruygrok, and the 19 Sansaporese were to proceed to Agats, a town down the coast, where the coastal steamer Oranje would pick them up on July 21st for the trip to Sorong and Biak.

Later in the day, we paid the paddlers with rice. Unlike the mountain people, the coastal Irianese had recently begun using cooking and eating utensils and had come to regard rice as a delicacy. We asked the village headman if a native dance could be arranged the following morning prior to our departure so that we could take pictures. The idea was well received. As we were having our evening meal small fires were started in the opening in front of our thatched dwelling. Shortly after dusk several men appeared with drums, hollowed-out logs about eight inches in diameter and two feet long. One end was open and the skin of some animal was stretched tightly over the other end. The small fires were used to temper the drums. The drummer would thump his instrument, hold it to the fire, thump again and repeat the procedure until he obtained the proper resonance.

This was to be a practice session for the next morning's performance. The drummers assembled in a tight group and as their monotonous thumping built up they were surrounded by an inner circle of men who started a shuffling movement in a clockwise direction and an outer circle of women who shuffled in the reverse direction.

Within an hour, the practice began turning into the real thing. As the tempo increased, the mood intensified, and everyone began to be carried away with the rhythm. The outer ring was joined by a group of nubile girls clad only in a spray of cassowary feathers tied firmly around their rumps.

We watched the scene in the flickering firelight until after midnight when we retired. The dance continued until shortly before dawn. When we left camp on the following morning, the entire village was still asleep.

After a long period of farewells, Del and I with our precious Ertsberg samples boarded the diesel launch and headed downstream for contact with the Beaver. The aircraft circled over the landing point on the river at exactly 1000. Suddenly I became quite excited. It must have been an outflowing of pent-up emotion that had accumulated over the past several weeks. For the first time I allowed myself to accept the thought that our expedition had been a success.

We greeted the pilot like a lost friend and rushed our cargo aboard. The plane lifted easily from the quiet tidal Koperapoka River. In about 30 minutes we were gliding in to a landing on the Mimika River opposite the small town of Kokonao.

The Mimika, like all other rivers in the region, was in flood. The pilot's objective was a 55-gallon steel drum, anchored in midriver but bobbing about like a cork in the fast flow of coffee-colored water. After landing in the river, the pilot skillfully adjusted his speed to the current so that he could leave the pilot's seat, run out on the left pontoon, snap a lead line to the hook in the oil drum, jump back into the pilot's seat, cut the prop and let the aircraft drift backward pulling against its mooring.

Dr. Feldman, the Dutch administrator, came out to our mooring at midstream in a small outboard to receive his mail and chat briefly with us. As the engine warmed up again, the pilot unhooked from the oil drum and allowed the aircraft to drift backward while he looked ahead for logs or trash. With nothing in sight, we took off after a surprisingly short run, crossed Wissel Lakes and landed on the northwest coast of the island at Nabire, the point from which some of our Carstensz air drops had originated and where we unloaded some mail and small cargo. An hour later we landed in the lagoon at Biak where John Bowenkamp and I had departed on the morning of May 29th.

Del Flint had a commitment to be at the International Geological Congress in Helsinki, Finland, within a few days, so I had arranged in advance for his departure from Biak on July 16th. Two days later, after repacking our samples for air freight to New York and completing a taped interview with the Dutch press for a joint radio program with the noted anthropologist Margaret Mead (then visiting in Hollandia), I was also on my way home.

Bob Hills and my wife were at Idlewild when I landed on July 25th. Both, I could tell, were quite concerned about my physical condition after a month and a half in the harsh wilds of New Guinea. I probably did look a little gaunt, but I felt terrific. I was in better shape than I had been for ten years and than I probably would ever be again.

Chapter Eight

In a little over six months, we had conceived and successfully completed an ambitious minerals exploration expedition to one of the most remote regions of the globe. During my flight back to New York in July, 1960, I was so exhilarated by what I had seen and sampled at the Ertsberg that I fully anticipated progress toward development would be expeditious. In a couple of years, road-building and mine construction could easily be well underway.

Instead, a couple of years later, I was doing nothing about the Ertsberg but spending lunch hours researching early explorations of Western New Guinea at the New York Public Library and evenings writing up my notes of the 1960 expedition, which I doubted would be of interest to anyone but my five daughters. The Ertsberg project now appeared dead and I had little hope it would ever be revived.

* * *

In the first weeks after my return to the States, everything had seemed to confirm my initial optimism. At the airport, Freeport president Bob Hills asked me to delay my vacation for a few days to report to the board of directors during their July 27th meeting at the company's midtown Manhattan headquarters. While I was careful at this meeting to detail the inconvenience of the deposit's location, I stressed that in my opinion it was the largest above-ground copper deposit ever found, that the grade of copper was high, and that we should proceed with further studies.

My remarks, I could tell, met a mixed reaction. Some directors picked up my enthusiasm. Others seemed to feel the Ertsberg could only be an expensive goose chase. In the minds of some of the skeptics, quite likely, was the fact that Fidel Castro, who had taken power in Cuba the year before, had just expropriated Freeport's new $120 million nickel project at Moa Bay and another at Nicaro, of which I had once been a general manager. To them, the time may not have seemed opportune for another expensive venture in a politically uncertain part of the world. In retrospect, those holding the negative view at that time logically should have prevailed. But they did not. I received authorization to go ahead with plans for Phase II: a more detailed investigation of the Ertsberg's ore grade and commercial potential.

The ore samples, air-freighted from Biak, were assayed at both our own and independent laboratories. The assays put the average copper content at a rich 3.5%. Metallurgical tests indicated that commercially proven techniques could achieve a high recovery of copper from the mined ore with no anticipated problems. Reviewing my survey data and Del Flint's geological report, mining consultants confirmed our estimates of 13 million tons of ore above ground and another 14 million below ground for each 100 meters of depth. Other consultants estimated that the cost of a plant to process 5,000 tons of ore a day would be around $60 million and that the cost of producing copper would be 16¢ a pound after credit for small amounts of gold and silver associated with the copper. At the time, copper was selling in world markets for around 35¢ a pound. From these data, Freeport's financial department calculated that the company could recover its investment in three years and then begin earning an attractive profit.

The calculation was sufficiently encouraging to justify moving ahead with the next part of Phase II: planning a substantially scaled-up ore sampling expedition. Our initial samples, which had been chipped from the surface of the Ertsberg and from talus on the ground, were useful for rough estimates. Before the company could make a final decision of whether to spend millions of dollars building a commercial facility, however, we needed to know much more about the character of the Ertsberg's ore. We would have to penetrate deep into the interior of the deposit with large diamond drills in order to obtain thick core samples. From these samples, we would be able to estimate the deposit's size and average metal content.

Diamond drills large enough to obtain core samples, though, weigh several tons. There was no way we could move drills in and core samples out with local porters and dugout canoes. I felt lucky merely to have gotten all the way to the Ertsberg with my geology pick and to have transported 600 pounds of ore chips back to the coast.

The only answer seemed to be helicopters. We had briefly considered helicopters for our first trip to the Ertsberg, though, and we knew they had serious weight limitations. The powerful turbine-powered choppers of today were just barely on the drawing boards. In 1960, the largest piston helicopters could lift only one passenger plus a cargo load of 250 pounds to an elevation of 12,000 feet. Carrying drills, core samples, people, supplies, and other equipment back and forth in 250-pound loads could take many months. The only drills that could be disassembled into small enough pieces would have very limited depth capacity, probably too limited for the studies we needed to conduct. Finally, no helicopter contracting company we contacted would even consider undertaking the job unless we first built emergency landing pads at two-mile intervals all the way from the coast to the Ertsberg. Installing these pads, we estimated, could take a year or more.

By the end of 1960, we had a program but no idea how it could be implemented. It seemed much like one of the physical impasses that plagued our overland expedition. Except that now we had no Moses Kelangin to come up with a solution.

The engineering quandary, meanwhile, was becoming compounded by political complications. The Dutch colonial rule over Western New Guinea was rapidly coming to an end.

Movements for Indonesian independence date back at least to the early part of the century. But they did not gather strength until World War II when occupying Japanese forces forced out the Dutch colonial administrators. After the Japanese surrender in 1945, Sukarno and Mohammad Hatta, the most prominent nationalists, proclaimed Indonesia an independent republic. Several years of often heavy fighting between Dutch and nationalist forces ensued. A United Nations commission was formed to resolve the dispute. Under heavy UN pressure, the Dutch in 1949 agreed to relinquish sovereignty over the former Netherlands East Indies. A new government was established and Sukarno was elected president.

The Dutch refused, though, to give up Netherlands New Guinea on the grounds that the islands had no geographic or ethnic connection with the rest of Indonesia. Public opinion in Holland had become very sensitive to the erosion of the country's once extensive colonial empire. And many Dutch religious groups, who had become active in New Guinea since the war, felt that bringing Christianity to the primitive natives was a holy mission that could not be relinquished. Throughout the 1950's, Indonesia attempted without success to force a change in Dutch policy. In late 1957, after failing to gain support for its cause in the UN General Assembly, Indonesia seized Dutch property and expelled most Dutch citizens from its territory. In late 1961, a band of paratroopers dispatched by Sukarno landed in Western New Guinea not far from the present access road to the Ertsberg and began a sporadic military offensive against Dutch forces.

Holland by now was not inclined to launch a strong

counteroffensive. Internal political support for retention of Netherlands New Guinea was waning. The government saw little point in a costly war thousands of miles from home. And international pressure on the Dutch to withdraw, particularly from the United States, was becoming intense. In August 1962, Holland accepted a plan by Ellsworth Bunker, who had been appointed mediator by UN Secretary General U Thant, to transfer administration of the territory to the UN with the understanding that it would pass from Dutch to Indonesian control in May 1963. Indonesia agreed to hold a referendum by 1969 that would enable the local populace to decide their ultimate allegiance. (In 1969, delegates selected to represent the local populace voted to remain part of Indonesia. The former Netherlands New Guinea was renamed Irian Barat, or West Irian, in 1963.)

The Dutch withdrawal put Freeport's position vis-à-vis the Ertsberg in something of a limbo, for it was the government of Holland that had authorized our exploration venture. In mid-1961 Freeport's engineering group concluded, since the development of sufficiently powerful helicopters seemed imminent, that the Ertsberg project should be pursued. But when we approached the UN in 1962, we were told that the UN was merely the caretaker of the territory and that we should discuss our plans with Indonesian officials. When we approached Indonesia through the country's embassy in Washington, we got no response.

Not long after Indonesia obtained control over Western New Guinea in 1963, then-President Sukarno, who had consolidated his executive power, made a series of moves which would have discouraged even the most eager prospective Western investor. He expropriated nearly all foreign investments in Indonesia. He ordered American agencies, including the Agency for International Development, to leave the country. He cultivated close ties with Communist China and with Indonesia's Communist Party, known as the PKI. And he withdrew Indonesia from the United Nations.

Certain that my first trip to the Carstenszweide had been my last, I became involved in some mining projects in Australia. Though not as grueling as scaling thousand-foot limestone cliffs, the work was demanding. During one year I flew around the world three times. I gradually forgot about the Ertsberg and concentrated on other ore bodies that, while less alluring and exotic, involved fewer headaches.

Early one evening in November 1965, I received a call at my home in Connecticut from Langbourne Williams, then chairman of Freeport. He asked me if it might not be the right time for another try at the Ertsberg project. Having spent much time closing my mind to that possibility, I was so startled I didn't know what to say.

I had only sketchily followed the recent political developments in Indonesia that occasioned Williams' call. During his years in power, Sukarno had worked hard with some success to achieve a measure of national unity, a rather formidable task considering that his nation consists of 13,000 islands scattered over 3,000 miles of ocean, and an incredibly diverse population that includes over 300 distinct ethnic groups and dozens of languages. But his regime had been marred by growing corruption, inefficiency, and injustice. Annual inflation soared past 600% and the economy was sliding toward bankruptcy. Internal uprisings against his rule had frequently occurred. Sukarno's own mental and physical health was said to be failing.

On the evening of September 30, 1965, members of the Communist PKI, hoping to exploit the swelling chaos, tried to seize power by assassinating six senior generals in the Indonesian army, whose leaders were anti-Communist. General Suharto, commander of the army's strategic reserve, quickly took charge of the army and crushed the insurrection. A wave of anti-Communism swept the largely Moslem country and the PKI and its front organizations were decimated. Sukarno, who was widely believed to have formed a secret alliance with the PKI and who had refused to condemn the leaders of the attempted coup, lost effective control of the government. Suharto's power grew. A veteran of the war for independence who later commanded the 1961 paratrooper campaign in Western New Guinea, Suharto and his new government began to reverse Sukarno's pro-Communist stance and his disastrous economic policies. In March 1967, the parliamentary assembly revoked Sukarno's lifetime mandate as president and appointed Suharto acting president. Sukarno was placed under house arrest where he remained until his death in 1970.

Indonesia's new economic policies seemed auspicious for Freeport. Not only was the new president an outspoken anti-Communist and economic pragmatist, but Williams had some even more encouraging private information which had been relayed by two executives of Texaco. Texaco had managed to preserve its large investment in Indonesia during the Sukarno regime largely through the efforts of its Indonesian managing director, Julius Tahija. An Indonesian from the island of Ambon, Tahija in World War II had served with distinction in the Dutch and Indonesian armies and had maintained close ties with the Indonesian government. From him, Freeport learned that the time was propitious to approach the new government about resuming the Ertsberg venture.

In a meeting with Tahija in New York in February 1966, Bob Hills and I assured him we remained enthusiastic about the project. In March, Tahija arranged a meeting in Amsterdam between us and General Ibnu Sutowo, Minister of Mines and Petroleum. Ibnu was well aware of the details of the 1960 expedition and Ertsberg's commercial potential. Indeed, word of our expedition apparently had spread widely in mining circles despite our attempts to maintain secrecy. Other concerns also had come to view the political changes in Indonesia as an opportunity to become involved in developing the deposit.

"People have been beating at my door," Ibnu told us. "Japanese interests are particularly active at high levels in an attempt to gain a foothold. But you people have already been at the Ertsberg and should be able to put it into production sooner than anyone else. I would like to welcome you back to Indonesia." His government, he said, would appoint a negotiating committee to discuss the project with Freeport.

Bob Hills and I consulted with Julius Tahija about whom

we should select as our advisor during the coming talks. At his recommendation, we enlisted Ali Budiardjo, who had filled several high government posts, including that of Secretary-General of Defense and Director of the National Development Planning Body during the 1950's. Having come to a parting of the ways with the Sukarno regime, he had taken leave of government service in 1959 to take an advanced degree at M.I.T. (His wife, who had served with the Indonesian Ministry of Foreign Affairs and who has a masters degree from Georgetown University, did graduate work at Harvard.) Both were members of highly respected Javanese families, and Budiardjo was in the elite group of Indonesians whom the Dutch educated and groomed for government service but who came to the forefront of the revolution against the Dutch. Members of this group have been running the nation ever since independence. A rather gentle, gracious, soft-spoken lawyer of impeccable morals who is perceptive and shrewd in his quiet way, Ali continues to enjoy the friendship and respect of officials throughout the Indonesian government. He took over from me as president of Freeport Indonesia, Incorporated, when I retired in 1974.

I arrived in Jakarta during June 1966 with several other Freeport executives for our first meeting with the government negotiating committee. We were among the first foreigners allowed into the country after the coup attempt. Soldiers were everywhere. At night tanks rumbled through the streets, which were otherwise deserted due to a strict 1900 curfew. We stayed at the huge Hotel Indonesia, one of Sukarno's many overly ambitious construction projects that had been rushed to completion in 1962 before the Asian games. Some 1,800 people were on the payroll, but on the evening of our arrival, the hotel had only 15 guests.

The meeting, held in a conference room at the Ministry of Mines, did not begin on a positive note. The committee had apparently not been given any new negotiating instructions for dealing with prospective mineral investments, and went by the contractual provisions of the "production-sharing agreements" developed by Indonesia in 1964 and 1965 under Sukarno for foreign oil contractors.

These terms required that: 1) Freeport would surrender title to the government of any equipment brought into Indonesia; 2) revenues, and particularly the foreign exchange, generated by the project would be controlled by the government; 3) project management would be in the hands of the government, with only technical direction by Freeport; 4) once Freeport had recovered its capital investment the Indonesian government would take over the project.

While some of these terms had been accepted by some oil companies whose fast cash-flow returns could justify such contract provisions, they were simply not suitable for the much longer lead times and heavier capital requirements of hard mineral investments.

When we showed our dissatisfaction with these terms, the committee members did not appear unfriendly. They seemed to understand the difficulties the conditions posed for us, and clearly wanted to do business with us. They just lacked the authority to take any other position. After three fruitless weeks, Anondo, the committee chairman, adjourned the meetings. He suggested that Freeport put forward a draft contract that it would be willing to accept. His committee would study the draft and reconvene the talks in three or four months.

Freeport was not looking for the old "concession" type of agreement which basically gave ownership of an area and its minerals to the contractor, but we felt there was middle ground in the less onerous "contract of work" type of agreement which the Indonesian government had used with oil companies prior to the production-sharing agreement. We forwarded in July a draft along these lines, drawn up by Freeport legal counsel Bob Duke.

In October, Duke, Nils Kindwall, also of Freeport, and I flew back to Jakarta for the second session. In view of the interest of other foreign companies in the Ertsberg, we had become apprehensive over the delay. Ali Budiardjo, though, had been effectively lobbying on our behalf. After our arrival, Ali took us to meet Saleh Bratanata, the new Minister of Mines for hard minerals who had assumed responsibility for hard minerals projects. Bratanata said he had studied our draft and hoped for an early agreement. He added he had received requests from the Dutch and Japanese ambassadors for permission for agencies in their countries to join with Freeport in developing the Ertsberg. But he assured us he had told the ambassadors, as he put it, "I am merely the traffic officer and Freeport is driving the car. As long as Freeport observes the rules of the road, they will continue to drive the car. If you want to hitch a ride, you will have to talk to Freeport."

That seemed reassuring, but we remained very concerned about just what the rules of the road would be.

When we reopened negotiations with the committee one Saturday morning in late October, we were quite unprepared for Anondo's opening remarks in which he agreed that we could retain title to equipment and material introduced in Indonesia. He further agreed that Freeport would have management control, and that funds generated from the sale of products could be deposited in banks outside Indonesia. He asked if we agreed.

When we could recover from our surprise, we said that, yes, we did agree. It was apparent that some efficient, high-level decision-making had gone on in the meantime which resulted in Anondo's getting clearance to do business with Freeport on the basis of a contract of work. Within a few days, we reached complete agreement except for certain tax and financial details. The Contract of Work would extend for 30 years. Freeport would not obtain title to the land, but it would, in effect, be the government's sole mining contractor for the project and would have exclusive mineral exploration and development rights for a 38-square-mile area around the Ertsberg. Agreement on the tax and financial details would have to await passage of a new Foreign Investment Law, which was to be presented to Parliament for approval in January. The new law, we were told, would completely reverse the foreign investment policies of the Sukarno regime.

At an elaborate press conference on October 31st, Anondo and I initialed a letter of agreement. While the agreement was subject to modifications by the forthcoming Foreign Investment Law, we now had about 75% of our desired work contract. We felt reasonably certain there would be no surprises in working out the remaining details.

The new law was duly enacted in January 1967. The government promptly returned expropriated properties to several companies such as Union Carbide and Goodyear. (In later years, the law would be responsible for attracting billions of dollars of foreign investment into the country.) In March, we began a third and we hoped final round of discussions with the negotiating committee. Besides Bob Duke and Nils Kindwall, the Freeport team included Bill Byrne, the company's tax expert, and H. C. "Pete" Petersen, president of Freeport's Australian subsidiary. As we sat down at the bargaining table, we were more than confident. We were even a little complacent. I should have remembered my lesson from my trek to the Ertsberg that things you think will be hard often turn out to be easy while things you are sure will be easy often turn out to be extremely difficult. Without warning, the committee announced a series of tax demands so onerous that financing the project would have been impossible. The government negotiators had toughened their position to avoid any criticism that the contract was too favorable to Freeport.

It took long hours of back and forth explanation and hard negotiating to reach an acceptable compromise that would give each side an equitable share in the project's revenues: for Indonesia because the project was in that country, and for Freeport because of the risks it was undertaking in the project. That the eventual contract was not a giveaway to the company is evidenced by the fact that in the first seven operating years of the project's life from 1973 through 1979, Freeport Indonesia was only able to pay two dividends to its shareholders, chiefly Freeport Minerals, totaling less than $15 million, while the invest-

ment at risk grew year by year to almost $300 million at the
end of 1980. Financing that investment, much of which
had been made in the early 1970's, involved substantial interest costs to the company. Considering all these factors,
it is evident that until very recently Freeport was getting a
very small return for its tremendous efforts. On the other
hand, the government continued to derive at least some
taxes every year from the project, even when the company
was in the red. Further, more than one thousand Indonesians had well-paid jobs whose payroll taxes also flowed
to the government.

Even after ironing out an agreement on the tax issue,
more trouble seemed to lie ahead. The Contract of Work,
including the tax compromise, had to be approved by
several ministries. Then it had to be presented for final legal
approval by the Praesidium, a high executive body that included President Suharto and the top members of his
cabinet. I was worried that it might become endlessly delayed by all these levels of bureaucratic review.

On March 30th, though, Ali introduced me to the Sultan
of Jogjakarta, Governor General of the Central Bank and
one of the Praesidium members. The Sultan and Ali were
old friends from the early days of Indonesian independence. After they exchanged a few stories about their es-
capes and escapades, the Sultan turned to me and said not
to worry. The signing, he thought, would be at the end of
the following week. I relayed the comment to New York.
Early on April 4th, Ali called me with the good news that
the Praesidium had approved our Contract of Work the
previous evening.

My spirits were high. But just an hour later Ali called

again. A crisis had arisen, and he told me to get to the office of General Suharto's private secretary as quickly as possible. The sharp and sudden swings in our fortunes, I must confess, were beginning to make me feel a little manic-depressive. I rushed to the building where I met Ali and several key members of the government's foreign investment technical advisory team, which had been formed to negotiate and supervise matters under the new Foreign Investment Law.

What had happened was that, on the previous day, when we thought the Praesidium had approved the contract, several ministers had strongly objected to certain legal wording in the document relating to approval procedures for the agreement. Ali and I put our heads together and decided that while the existing language in the agreement had been reached only after painstaking compromise, the demanded change was of less importance than attaining the goal of a signed agreement. I knew that the company's lawyers would not like the change, but exercising my prerogative as the man on the spot, I took out my pen and made the necessary changes.

Late that afternoon, Bob Hills arrived from New York. En route from the airport, I told him of the day's ups and downs. Signing, we had been told, would take place first thing in the morning. The morning, though, came and went. No one seemed to know what, if anything, was going on.

By the next morning, we had become extremely anxious. It was a little like waiting for bearers to show up. There didn't seem to be anything we could do but wait and hope. Late in the afternoon we were advised that, in fact, all was well and the signing would occur at nine the following morning at Minister Bratanata's office, which happened to be the former residence of the manager of Shell Oil.

Bob and I arrived at 0845. A table had been set up for the signing of eight documents, four for Indonesia and four for Freeport. Young ladies beautifully costumed in native attire were present to pass the documents from hand to hand. Television cameras and lights were moved into place and members of the press jockeyed for favorable positions. A few minutes before nine, a group of top Indonesian officials arrived together with Marshall Green, the American ambassador, who had given us valuable advice during the negotiations.

Promptly at 0900 Minister Bratanata moved to the center of the table. Bob took a position at his left. I occupied the right-hand seat and Ambassador Green moved in behind me to make some amusing and encouraging remarks as we signed the documents, all embellished with colorful ribbons and seals. This was the first Contract of Work to be signed by a foreign mining company under the terms of the Foreign Investment Law of 1967.

Lunch followed at the residence of Ambassador Green for the Sultan, four members of the Cabinet, U.S. Embassy personnel, and Hills, Budiardjo, and myself. That evening Bob and I were hosts at a dinner for 22, including Ambassador Green, Indonesian government officials, and all of the members of the Negotiating Committee.

The elaborate celebrations seemed an appropriate end to six years of inaction and an appropriate start to the Ertsberg project's long-awaited Phase II, an intensive and detailed analysis of the Ertsberg's ore.

Chapter Nine

To his friends and enemies, Ted Fitzgerald, a tough, wiry Australian, was known as Ever Ready Ted, a nickname derived from his propensity for engaging in a fight with virtually anyone at the least provocation. Most of Fitzgerald's fighting, though, took place at such standard locales as bars. As he neared the south coast of New Guinea early in May 1967 in his converted World War II LTC, he certainly wasn't looking for trouble. The 60-foot vessel, which had left Darwin seven days earlier, was loaded with several tons of equipment, including a bulldozer, a truck, and diesel generators. These items had been assembled by Pete Petersen and Reg Barden, another Freeport man in Australia who was accompanying Fitzgerald, for Phase II of the Ertsberg project.

About five miles off the coast, Ever Ready Ted spotted an unidentified craft approaching at high speed. At a range of a few hundred yards, it fired a warning shot across the LTC's bow and raised signals directing it to stop. Outgunned, Fitzgerald turned his ship toward the other craft, which he could see now belonged to the Indonesian navy coastal patrol. In a bit of unfortunate seamanship, Fitzgerald managed to sideswipe the naval vessel and tear off a large piece of its cowling. He and his crew were placed under immediate arrest for invading Indonesian waters and escorted to the coastal town of Kokonao, our radio base for the 1960 expedition.

Because of the shallow waters off Kokonao, both ships dropped anchor and the coastal patrol officers and their captives transferred to two small outboards for the trip to shore. Since the Indonesians spoke no English and the Westerners no Indonesian, communication was limited to sign language. Ever Ready Ted was not quite sure how or if the incident could be resolved.

In Kokonao, however, Reg Barden enlisted the aid of Margaret Smith, a Protestant missionary who spoke perfect Indonesian. Radio contact was established with Indonesian naval authorities who assured the coastal patrol officers that, though nobody had gotten around to advising them of the fact, Fitzgerald's ship indeed had clearance to enter Indonesian waters.

Fitzgerald, Barden, and the crew returned to the LTC, weighed anchor, and proceeded with their first assignment: scouting for a suitable location for a coastal helicopter and supply base. Helicopter technology had improved since 1962. Current models had a lift capacity at 12,000 feet of over 1,000 pounds—fully adequate for the job of moving diamond drills, personnel, and other equipment to the Ertsberg. My suggestions for a base had been Timika, a coastal village not far from Kokonao where the Japanese had constructed a small airstrip during the war and where Kroonduif Airlines more recently had landed DC-3's. The strip was still pockmarked from bombs dropped by the Australian air force. After several days of cruising along the coast, Barden concurred with the idea.

The LTC unloaded its equipment at Timika and the bulldozer began clearing the heavy vegetation along the shore and preparing foundations for buildings. Over the next few months, Fitzgerald brought in several additional boatloads of equipment. The cargo included duplicate sets of prefabricated buildings from Australian suppliers—one for the coast and one for the Ertsberg—consisting of a mess hall, kitchen, and dormitory for 30 people. Also imported were two completely disassembled Bell 204 helicopters acquired from Petroleum Helicopters in Louisiana. Three helicopter pilots and two mechanics arrived in August along with Balfour Darnell, an old Colombian gold-mining associate, the best man at my wedding in 1940, and an individual with a tremendous affinity for machines of all kinds whom I had convinced to give up a retirement of golf and fishing to head up our diamond-drilling operation.

Putting the two helicopters together with only the most rudimentary supporting equipment was a delicate and cumbersome task. Each day, an intent crowd of coastal Irianese gathered on the airstrip as the mechanics opened boxes, took out parts, and fitted the pieces together. The spectators, who had never seen a helicopter, wondered about what might be emerging from the growing metal assemblage.

The first craft was ready to fly on the morning of September 20. Don LaFreniere, the chief pilot, climbed up a small ladder, stepped inside, and closed the door. The Irianese let out a loud wail. They were certain the metal monster had gobbled him up. As the blades spun and the chopper lifted into the air, they fled in fright into the nearby mangroves.

Bal Darnell was working on the logistics of the Timika-Ertsberg shuttle. Gravel bars in the Ajkwa and Otomona Rivers were scouted for their suitability as an intermediate landing pad between the coast and the mountains. He selected a site on the Ajkwa about five miles upstream from the confluence of the two rivers. Though its elevation was only 300 feet, the spot was equidistant between the sea-coast camp and the Ertsberg camp at 11,600 feet. During the next three years, this pad was used hundreds of times, most often to store supplies brought up from Timika while the pilot waited for the weather at the Ertsberg to clear.

The Ertsberg weather, indeed, was a continual impediment throughout Phase II. Helicopters cannot be flown safely at night or in severely reduced visibility. Landing at the Ertsberg was possible only early in the morning before the inevitable clouds invaded from the lowlands. Often the Ertsberg would be completely socked in for two or three days at a time. Even more often, clouds would move in so soon after sunrise that a helicopter couldn't make it up there in time.

Not until early October was a helicopter able to make the first landing on the soft ground on the west side of the Ertsberg. Yet the pilot had no time to survey the Carstenszweide. Clouds were beginning to fill up the canyon between the mountains through which the flight path back to the coast led. He was forced to make a hasty retreat.

Conditions were better the following morning. Both craft were able to drop off loads of two-inch planks for the construction of a permanent landing pad. The next day they brought in three canvas tents and kitchen equipment but early cloud arrival prevented any delivery of personnel and progress on camp construction. It took us two weeks for eight men and living facilities to be flown to the site.

During one early morning landing, the pilot noticed that sticks of light-colored wood one or two feet high in the shape of a cross had been stuck in the ground in a ring around the Ertsberg. Bal Darnell found out from missionaries that these were *saleps* or "hex sticks." They indicated the local Irianese had a decidedly unfriendly attitude about what was going on. Bal radioed me in Darwin, Freeport's Australian headquarters, where I was making supply arrangements, and asked me what he should do. I had never heard of hex sticks before. My only suggestion was to find our old problem-solver, Moses Kelangin.

Bal located Moses in a village 30 miles east of the Ertsberg called Akimuga where he was the *kepala* or chief. Moses agreed to take a helicopter to Wa, the small village near our old Base Camp and ten miles south of the Ertsberg. Wa's residents seemed the most likely source of the hex sticks.

The helicopter landed in the rocky bed of the Aghawagong River. The village was deserted. Everyone, apparently, had gone into hiding at the approach of the aircraft. After repeated calling by Moses, a small boy appeared from the bushes. Moses told the boy to ask the others to come out for a talk. Gradually, more villagers began drifting into town. They told Moses the area around the Ertsberg was sacred. Moses explained that the white men were not trying to appropriate the villagers' land. They just wanted to test the rocks. As evidence of their good faith, Moses said, the white men would give the villages gifts of food and other items. Peace was reestablished, though, as it turned out, only temporarily.

Del Flint and I arrived in Timika from Darwin on November 7 aboard the Edwina Mae, a single-screw, 40-foot shrimp boat which made the 700-mile run in five days. The Arafura Sea, which often has violent storms, as we would later appreciate, was on its best behavior. We enjoyed a calm, glassy sea for the entire trip, which was fortunate since we had a deck cargo of 175 drums of diesel and aircraft fuel and very little freeboard.

We flew up to the Ertsberg the following day. It was a peculiar experience for me to have the drudging 17-day trip reduced to an effortless 45 minutes. The Carstenszweide was losing some of its remote, other-worldly atmosphere. Fabricated buildings were being erected and a 25-horsepower diesel generator was being installed.

After we located the most appropriate spots on the Ertsberg for the diamond drills, I returned to the U.S. Del Flint remained to assist Bal and his assistant John Currie in starting up the diamond drills and training several Indonesian geologists to log drill cores and prepare samples.

Three large diesel-powered drills made by Canadian Longyear, broken down into small components for helicopter transport, arrived on December 1st aboard Fitzgerald's LTC along with two mechanics. A week later, 24 Canadian diamond drillers were brought in. Five days later they had settled into the completed buildings in the Carstenszweide and the drilling program was underway. The drills, whose tips are rings of hundreds of tiny black industrial diamonds, began extracting cores an inch in diameter from holes as much as 1,200 feet deep into the Ertsberg.

After two months of seven-day-a-week, 24-hour-a-day drilling, we obtained 25,000 feet of drill core. By previous Ertsberg standards, the operation was racing along with uncommon speed, efficiency, and ease.

While the drilling continued, I began working on Phase III, the preparation of an elaborately detailed commercial feasibility report. It would consist of a complete engineering plan for the eventual production facilities, estimates of their construction and operating costs, and extensive financial forecasts of the market for the mine's output. On the basis of the feasibility report, Freeport would make its final decision on whether to go ahead with Phase IV—construction of the production facilities—and various lenders would decide whether to make available to the company the necessary funds.

I flew to San Francisco to talk with executives at Bechtel Corporation, the huge engineering and construction firm that has built industrial facilities throughout the world. We wanted the company to supervise preparation of the engineering study. As a first step, Bechtel agreed to discuss forming a joint venture with J. H. Pomeroy Co. (now Santa Fe-Pomeroy, Inc.), a smaller firm specializing in marine installations, to prepare initial conceptual engineering studies. In February, representatives of the two firms rendezvoused with me and Pete Petersen in Darwin. We crossed over to Timika aboard the Sundowner, a 50-foot, twin-engine launch.

Though they had been all over the world, the Bechtel and Pomeroy people were nearly overwhelmed by the formidable construction problems. Their awe and amazement, I was coming to realize, was common among people seeing the area for the first time. No matter how much is described about it beforehand, no matter how many pictures are seen, it is just impossible until one actually gets there to appreciate just how harsh and obstructive the terrain really is. For 14 years, I gave my wife elaborate accounts and showed her numerous movies, but when Ann made her first trip to the Ertsberg in 1974, she was still absolutely dumbfounded. After five days of investigation by foot, dugout canoe, and helicopter, the Bechtel and Pomeroy representatives regained their bearings and agreed to begin the necessary engineering data-gathering as soon as Freeport gave the go-ahead.

When we reached Timika for our return voyage to Darwin aboard the Sundowner, a severe storm was raging.

The sea was being whipped by gale force winds and we delayed our departure for 24 hours. The winds had not abated by the next afternoon, but the skipper decided to go ahead anyway. At 1900, we rowed out to the ship, which was anchored in the Timika River. The plan was to sail around midnight on the high tide. Trying to get some sleep while we could, we settled into our bunks. Opening my eyes the next morning, I found myself looking not at the open sea but mangrove trees. The skipper had changed his mind. And during the night, the Sundowner had dragged its anchor and backed up the river for nearly a mile.

At 1700 on the third day, despite the persisting gale, we tried again. Bal Darnell, who watched us leave with binoculars from the shore, later told me we rolled 45 degrees when we first hit the open sea and headed directly into the wind and waves. Actually it was even worse than that. I was foolishly standing at the rail when the first large wave hit us. It broke completely over my head and landed on the other side of the ship. For the next 12 hours, we were in continuous violent rolls. From our departure until 2000 that night we stood up in the cabin desperately clutching the uprights of our bunks.

Eventually Pete Petersen and I crawled into our upper bunks, where we managed to secure ourselves by bracing our feet on the foot board and wrapping one arm around a stanchion. About every fourth roll the ship would plunge forward into a wave, sending a torrent of water cascading over the top deck. A few quarts would dribble into the upper bunks. Fortunately, the water was warm.

When we awoke in the morning we were still heading directly into wind and waves. Sometime during the night a ventilator had been torn from the forward deck leaving a ragged hole that had been stuffed with burlap bags. Some planking on the starboard bow had also been ripped off and a part of the windscreen in the pilot house had been smashed. Late in the afternoon we came under the protective lee of the Aru Islands in the middle of the Arafura Sea, which for eight hours gave us some relief from high seas.

About midnight, we passed the southern tip of the Arus and altered course to the southeast toward Darwin. The respite quickly ended and we were caught again in the full force of the storm.

The first things I noticed after I awoke the next morning were that there was no sound from the engines and that we were drifting downwind toward the east and away from Darwin. On deck, I found Pete making a very close examination of the ship's one small and totally inadequate life raft. "We're in trouble," he said in a surprisingly calm voice. "The main pump is down and submerged. We have engine problems and the engines have been stopped. The water level is within six inches of killing the engines if they are started. The emergency pump is fouled and they can't seem to get it running. We just received a radio message that there is a cyclone off Darwin and all small craft are advised to seek shelter immediately."

Just as he finished talking I heard the stuttering sound of a motor trying to get itself going. The single cylinder emergency pump was finally working. A crew member rushed top side and peered over the side as a great gout of garbage, bits of waste, pieces of rubber and other trash spurted out of the pump discharge. After two hours the water level was lowered, the main pump was repaired, the engines were started and we returned to a southeasterly course.

The engines, though, were still running very hot and had to be shut down about every four hours for special lubrication. During the down time, the ship had to be pointed downwind and away from our target. Early on the fifth day we sighted the north coast of Australia but we were 150 miles east of Cape Don near Darwin. It took 18 hours of sailing dead into the wind to reach the passage leading to Darwin. When we arrived off Cape Don the tide was running out at full force. For two hours we just barely held our own in the swirling current. At 2000 on the sixth day, wet, weary, and hungry, we at last dropped anchor.

About a year later, on a return trip from Timika to Darwin, the Sundowner hit a large log and sank off one of the small islands on the south end of the Aru group. Fortunately, crew and passengers reached shore with one rifle and limited supplies. They were able to subsist on venison and oysters until they were rescued two weeks later.

At the end of February, 1968, we received an analysis of the first drill cores from the upper section of the Ertsberg. They indicated ore of more than 3% copper. But as results came in from deeper holes in the lower part of the deposit, the average grade of the whole mass dropped to 2.5%. This is still quite rich by copper-mining standards. Many commercial mines have less than 1% copper. Most commercial mines, though, are more conveniently located. And the new figure was much lower than the close to 3.5% average of our surface samples obtained in 1960. The total amount of ore was estimated, further, at 33 million tons against original estimates of 50 million tons or more.

These results gave rise to a wholly new set of problems that came as close to killing the Ertsberg project as did our earlier troubles with the New Guinea terrain and Indonesian politics. To keep the project alive, I found myself transformed from a mining geologist into a salesman.

When I had asked for $120,000 to finance the 1960 expedition, the reaction by Freeport's executives and directors was unanimously enthusiastic. When I asked to proceed with Phase II, there was mild opposition, but supporters of the project easily prevailed. Now, with the ore grade shown to be only 2.5%, the opposition grew more intense. Phase III, preparation of the feasibility study, would cost about $2 million. Several directors and officers thought it could be a waste of money. And several other felt Freeport should not go ahead unless it could find a partner to share the risk, even though this would require giving away half of the company's interest in the deposit.

The opponents, it must be emphasized, were well-intentioned. They believed they were acting in the best interests of Freeport just as much as I and the other supporters did. Given the information available at the time, a disinterested but informed observer could just as easily have sided with them as with us.

To understand the debate that raged throughout 1968, one must appreciate that historically Freeport had been a conservative company. It had almost never borrowed money. With one or two notable exceptions, it had not embarked on expensive new projects that were known to involve large risks. When Freeport had taken on such projects, it had often sought out partners to share those risks.

Over the years, this attitude paid off. Freeport was a solid moneymaker and its traditional sulphur business was very profitable. The company was in much better shape than some other mining firms that had taken larger risks and suffered large losses. Opponents of the Ertsberg project continued to remember well Castro's expropriation of our Cuban nickel plants, which had been one of the largest and riskiest projects in Freeport's history. Freeport, clearly, was no stranger to political risks.

Skepticism about the Ertsberg also derived from its variance with Freeport's usual mining experience. Sulphur is a sedimentary deposit. These deposits occur in large, flat beds whose thickness is relatively uniform and whose size can be measured quite precisely. One can usually predict at the outset within an error of only 10% or 20% what the eventual yield will be. Igneous deposits like the Ertsberg, with which most Freeport executives had had little direct experience, are more difficult to measure. Their veins tend to extend deep underground and have unpredictable configurations and degrees of richness. One usually doesn't know precisely how much ore a deposit contains until the mine is close to being played out. I argued that though we were only certain at this point of 33 million tons, the eventual tonnage of the deposit could be much greater. For perfectly valid reasons—for I could easily have been wrong—the opponents felt I was overly optimistic. They did not want to commit a large investment to a deposit whose size was so uncertain.

To counter this view, I dispatched Del Flint back to the Ertsberg in April 1968. He was to seek evidence of additional mineral possibilities near the Ertsberg. We were especially interested in the stains of green malachite that Gus and I had spotted in nearby limestone cliffs in 1960. During two days of exploration, Del found strong surface expressions of copper over an area 3,000 feet long and 1,000 feet wide within a mile of the Ertsberg. And he spotted a new outcrop of Ertsberg-type ore at an elevation of 14,000 feet.

But he also had bad news to report. By the time he had arrived, the diamond drilling team had been gone for several weeks. While the camp was unattended, local Irianese had broken in, destroyed a number of the buildings, stolen supplies, and left the area in a shambles. Their motive, apparently, was not hostility. As in the 1960 Base Camp incident, it was just that the goods were there for the taking. We later questioned villagers in Wa who, in return for food and items of clothing, had promised to watch the camp while we were gone. They blamed other tribes, but looted items were visible around their village.

Despite Del's report of additional copper deposits, Freeport's senior managers decided to close the camp and, pending further study, bring the equipment and helicopters back to the United States. I received a letter from one executive praising my initiative and courage but indicating that the Ertsberg was obviously dead and suggesting that I go back to Australia and try to find a good nickel mine.

Instead, I spent several months drawing up a detailed proposal for Phase III. In October, I was invited to submit it at a morning meeting of the executive committee. The meeting, from my standpoint, did not start well. I was given the opportunity to defend my position. But the general consensus was hesitant and even negative. I was very disheartened. I began to feel I had no choice but to resign from Freeport.

Then Augustus C. Long, a board member and for many years chief executive of Texaco, took the floor. He noted that the market for sulphur, which accounted for 85% of Freeport's sales, was weak. At the time the company did not have any major projects other than the Ertsberg on the horizon with much economic potential. "If we don't give Wilson what he's asking for," Long concluded, "we as directors might be derelict in our duty. The project has high risks but it promises good rewards." After a brief silence, he moved that the $2 million necessary for Phase III be authorized. The motion passed unanimously.

As I was walking out of the boardroom, Long turned to me. "You'll have to buy me lunch for that," he said with a smile. I would have bought him lunch every day for a month, but I didn't have the time. In a few days, I left for San Francisco to tell Bechtel and Pomeroy to go ahead. And I arranged for the helicopters and equipment to be shipped back to Timika and our on-site staff remobilized.

A major part of the engineering studies involved metallurgical tests to simulate as closely as possible actual commercial operation. These tests require a large amount of ore. The reason is that the precise composition of different ore deposits varies greatly. A mill must be precisely designed for the particular characteristics of the ore it handles. Different kinds of ore require different processing techniques. Ideally, a 10,000-ton bulk sample is the minimum needed to permit operation of a small-scale test facility to be representative of a full-scale commercial plant and to ensure that any flaws in the treatment process are discovered and corrected before the commercial mill is built.

Transporting 10,000 tons of ore from the Ertsberg to the coast by helicopter was impossible. At 1,000 pounds a load, it would have required nearly 20,000 separate trips and perhaps ten years of daily shuttle flying. The maximum we felt we had the time and money to move was only 300 tons. We decided to see if we could get by with it. If the Ertsberg ore had later turned out to have had very unusual characteristics requiring special processing, the deposit probably never would have been mined, but our gamble worked. The ore proved very amenable to standard treatment methods and the mill test on the 300 tons of bulk samples gave us acceptably accurate results.

Yet even those 300 tons were not obtained without considerable effort. The only people we could recruit on short notice were a tough Australian miner and a crew of ten young shrimp fishermen from Darwin who had never before seen a pneumatic drill or a stick of dynamite. Nevertheless, they performed extremely well. In a month, they drove a small 6 x 4-foot adit 300 feet into the side of the Ertsberg to obtain the required sample of ore.

The bulk samples were loaded into 656 55-gallon oil drums, each of which weighed close to the lifting limit of the helicopters. Because of the usual abominable weather, on many days we were not able to move more than two drums. On five days, we couldn't move any drums at all. The helicopter airlift of 300 tons of ore samples took nearly five months. Another two months was required to move the 656 drums by LTC from Timika to Darwin where they were transferred to ocean-going ships for shipment to Vancouver, Canada. From there, they moved by rail to the Michigan Technological Institute at Houghton, Michigan.

As the Bechtel-Pomeroy engineering studies began, another less serious transportation dilemma developed. The Bechtel-Pomeroy staff on the project ranged between 30 and 50. Outside consultants and specialists were continually coming in and leaving the job site. The traffic between Timika and Darwin became too heavy for the available fleet of LTC's and shrimp boats in Darwin. The crossing (five days for shrimp boats, seven to eight for LTC's) was time-consuming, uncomfortable, and sometimes dangerous. We had to have an air shuttle. But the airstrip at Timika, along with a large section of the coastline, had sunk between 10 and 15 feet since 1960. At high tide, about 800 feet of the strip was covered by water flooding up through old bomb craters. What we needed was an amphibious plane that could land on the Timika River. No one, however, was then making long-range seaplanes since they had little commercial value. Our objective became a secondhand survivor of the fleet of Navy PBY's that had seen extensive action during World War II.

In a search that took him from South America to Alaska to Malaysia, Dick NcNally, Freeport's chief pilot, ran across several candidates, but all were too worn and weary for a new tour of heavy duty. One day, Dick called me from San Francisco. By coincidence, in my office at that moment was an Australian who had recently purchased, and said he might be willing to sell for $295,000, a PBY that was located in a hanger across the bay from San Francisco in Oakland. I told Dick to hurry over and take a look.

Dick called me back two hours later and told me to make a deal at the asking price. Although that was in excess of my budget, the plane turned out to be in remarkable condition and perhaps the last of its kind in the world capable of meeting our needs. Delivered to the Navy after the end of the war, it had been purchased as war surplus by Howard Hughes. Hughes replaced the machine gun mounts with large plastic bubbles, fitted the interior in executive style, installed hydraulic equipment to raise and lower small boats that fitted under each wing, and put in housings on the upper wing surface to accommodate swivel chairs. He intended to use the craft as a flying boat for salmon fishing in Alaska. Like Hughes's larger and better-known wooden "Spruce Goose" flying boat in Long Beach, California, the PBY had sat unused at the Oakland airport for more than 20 years.

The airframe was absolutely sound. It required no change other than the elimination of the small boats under the wings, the swivel chairs, and incidental fishing gear. Though the two engines had practically zero hours, they had deteriorated over the years and had to be replaced with new units. We hired a crew to fly the plane out to Darwin. The big lumbering craft island-hopped across the Atlantic without difficulty, but it was grounded in Teheran by snowstorms, grounded in India by bureaucratic procedures, and delayed another 10 days in Malaysia because of mechanical problems. It arrived in Darwin nearly two months behind schedule. The long journey apparently sapped the morale of the crew. In Darwin they made it clear they had negligible interest in doing any flying to Western New Guinea. As a temporary replacement, we dispatched Freeport pilot Bob Stiles, who made the first landing in Timika after a five-hour flight in March 1969. Bob continued for two months on a twice-a-week schedule while we searched for a permanent pilot.

We located Harry Swanton, a large, burly Alaskan bush pilot of the sort whose feats of derring-do quickly turn into legends. Keeping the craft in the air with spare parts from an old World War II dump he found in Biak, Swanton made the Darwin-Timika run until March 1971, when a 6,000-foot runway suitable for land-based planes was carved out of the jungle near the access road. We sold the PBY to an adventurer who took off for what he said was an around-the-world flight. Swanton, meanwhile, went off to another assignment in Malaysia.

For eight months, the Bechtel-Pomeroy engineering staff checked ocean currents near offshore landing sites, surveyed the Timika River channel, drilled holes and made soil tests for the proposed plant and port sites, measured river flows, collected weather data, and attended to a thousand other details. The first engineering results were ready in August. They were based on the concept of mining copper and iron ore by open pit and transporting the ore to a port facility in slurry form. In other words, the ore would be crushed near the mine. The crushed ore would be mixed with water and pumped down to the coast through a 12-inch pipeline. This would be cheaper and more efficient than hauling the ore by truck. On the coast a concentrator would pulverize the ore into a fine powder in order to separate from the waste rock a concentrate of copper, small amounts of associated gold and silver, and another concentrate of iron. Bechtel-Pomeroy put the capital cost of the facility at $160 million.

In the view even of the Ertsberg's most enthusiastic supporters, that cost was beyond Freeport's financial capabilities. We asked Bechtel-Pomeroy to make a new study of a less ambitious facility. Though the Ertsberg was about 40% iron, recovering the iron, in our view, was not commercially attractive. Pig iron was (and is) in plentiful supply worldwide and sells currently for only about $200 a ton versus $1,600 a ton for copper. The revised mill, we told Bechtel-Pomeroy, should not recover iron. In the new design submitted by Bechtel-Pomeroy, the concentrator was located in a canyon at the base of the 2,000-foot headwall leading up to the Carstenszweide. The copper concentrate, which would be about 32% copper, would travel to the coast through a much smaller slurry pipeline. In September, we received a preliminary report that the revision was commercially feasible. It would be less expensive: $100 to $120 million. Bechtel-Pomeroy promised to have the full formal feasibility study ready in December.

Freeport, meanwhile, was lining up the necessary financing. Money for a project of this size usually must be obtained from many sources, as no one lender would want or be able to bear the full risk. Paul Douglas, son of the late distinguished United States senator from Illinois, then head of Freeport's international sulphur operations and now its chief executive officer, spent much of 1969 obtaining long-term sales commitments from copper smelters in Japan and Germany. Nils Kindwall, then treasurer and now senior vice-president and chief financial officer, circled the globe seeking loan funds from domestic and international financial institutions.

The Ertsberg's inaccessibility had complicated every other aspect of the project, and the financing campaign was no exception. The less routine a project is, the more nervous lenders become—and with good reason. At one point, I made a presentation about our access road plan to a senior engineer at the Export-Import Bank in Washington. The Ex-Im Bank is a federal banking corporation that we hoped would guarantee loans by some American banks against political risks. Its support for the project was crucial. The engineer from the bank was clearly disturbed by the fact that at the time we had only a rough idea of where the road was going to be built. I explained that while we had flown over the proposed route many times, the ruggedness of the terrain had made it impossible for us to conduct a detailed ground survey. After each day's advance, I said, we would have to stop and figure out just how best to proceed the next day. And that, in fact, was how the road was built. The engineer, though, said we would have to supply a detailed plot of the road plan or we wouldn't get the guarantee. The Export-Import Bank, he said, does not guarantee loans for projects constructed on a play-it-by-ear basis. We were able to convince the engineer's superiors that the Ertsberg justified an exception to their normal procedures.

By September, after many weeks of involved negotiations, Douglas and Kindwall were well on their way toward completing an intricate $100 million loan package that ultimately included some 30 parties from three countries. Among the pieces:

• $20 million from five major Japanese trading companies and eight smelting companies. The smelting companies agreed to purchase at least 65% of the Ertsberg production for a 15-year period.

• 80.5 million Deutsche Marks (the equivalent of $22 million) from Kreditanstalt für Wiederaufbau, the German development bank, along with a contract with Norddeutsche Affinerie in Hamburg, the largest copper smelter in Europe, to purchase the balance of the Ertsberg's production for the life of the mine.

• $40 million from seven large American banks guaranteed by the United States Export-Import bank.

All the lenders agreed to provide additional funds up to 20% to cover cost overruns. In addition, Freeport Minerals (or Freeport Sulphur Company, as it was then called) was required to commit to the project $20 million of its own capital plus 20% for overruns. Beyond this, however, Freeport Minerals would not be responsible for any losses by Freeport Indonesia, the Freeport subsidiary that would develop the Ertsberg.

The package had a number of important contingencies. It was subject to acceptance of the final feasibility study by all parties and to approval by several government agencies in the United States, Indonesia, Germany, and Japan. And if one member of the financing group dropped out for any reason, everyone else would have the same right.

By late September, the Bechtel-Pomeroy engineers had returned from Timika to San Francisco to prepare the feasibility report. Only a few Freeport personnel remained at the Ertsberg. There was nothing for anyone to do but wait until December.

I suspected, accurately as it turned out, that a final go-ahead might not come for several months after the feasibility report had been received. Something had to be done in the meantime, I felt, to maintain the project's forward momentum, both at the job site and in New York. Once lost, momentum is hard to regain.

In early October, I proposed to the Freeport management and board that $300,000 be appropriated to begin preliminary work on the access road, which was to be the first part of Phase IV. The reaction was not very positive. Some who had consented to Phase III with reluctance now felt that further expenditures should be delayed until the feasibility report was approved and the financing assured.

Again the decisive voice came from one of the company's outside directors. Admiral Arleigh Burke, former

Chief of Naval Operations, argued strongly and persuasively for maintaining the momentum and risking Freeport's funds. The vote in favor was unanimous. With the $300,000, we purchased four small bulldozers in Singapore and rented six trucks in Australia. The equipment was shipped to Timika and pioneer road-building began.

At 2300 on December 19th, a courier from San Francisco arrived at my room at the Yale Club in Manhattan with 25 copies of the engineering report. Bechtel estimated the cost of the facilities, including provisions for escalation, contingencies, and a $3.4 million fee, at $90.5 million. To that figure, Freeport added $13.5 million for interest incurred during construction; $11.75 million for overhead, development, and preoperating costs ($6 million of which had already been spent); and $4.5 million in working capital. The overall estimated cost was $120 million. (The eventual cost, due to inflation, engineering modifications, and such other factors as the incredible difficulties of road-building turned out to be $200 million.) Bechtel-Pomeroy estimated that the construction would take three years: two years for the access road and one year for the mill and other production facilities once the road was in place.

A special meeting of the board of directors was called the next day to consider the report. As I expected, it voted to defer a final decision until the report had been approved by the lenders. In a subsequent board meeting in January, the board voted approval in principle subject to the actual signing of all required financing and regulatory agreements. I was authorized to sign a letter of intent with Bechtel-Pomeroy, whom we agreed should be given the construction contract, to enable them to begin mobilizing men and equipment.

As had happened so many times in the past, unforeseen obstacles kept cropping up all spring. Approval was delayed from boards and agencies in Germany and Japan who met only at infrequent intervals. The Japanese agencies were especially slow to act. Freeport had to resort to a short-term bridge loan of $13 million to keep the money flowing. In late May 1970, the final Japanese agency issued its approval, in June Freeport drew down the first portion of the loan funds, and at last, almost exactly ten years after I took my first look at the fog-shrouded Ertsberg, we were able to proceed with certainty on Phase IV.

Chapter Ten

Early one morning in 1971, Steele Ansley was standing on a completed section of the access road through the swampy regions near the south coast of West Irian. A Freeport supervisor, Ansley was making calculations on the progress of the road construction. Progress had been slow. First mangrove trees and other surface growth had to be cleared with chain saws. To establish a firm foundation, a bottom layer of logs and then load after load of gravel from nearby riverbeds had to be dumped on the soft, Jello-like muck composed of mud and dead vegetation. The muck was often 12 feet thick or more. The ground was so unsteady that when a load of gravel was dumped, trees on the sides of the road would quiver and shake. Sometimes 50 cubic yards of gravel, eight to ten truck loads, would be required to advance the road a single foot. Sometimes the daily advance would be only 25 feet.

Ansley had no trouble understanding why progress had been slow. There was one thing, though, that he was having considerable trouble understanding. In fact, it had him baffled. At the end of each day, Ansley's men would measure the additional footage that had been laid down. After several weeks of work, these daily figures, when added together, equaled nearly twice as much distance as the length of the road that was actually in place. Close to two miles of road had been laid down. But the actual road was now only a mile long. Glancing at where the previous day's construction had ended, which was about 100 feet from where he was standing, Ansley tried to figure out what had happened to the unaccounted-for road.

Rather graphically, the answer abruptly presented itself. "Before my very eyes," he recalls, "the last 75 feet or so of the road broke off and started sinking." In less than a minute, the sinking section had disappeared beneath the muck. The missing two miles of road lay underneath the visible two miles. The road, in effect, was requiring a double layer of fill.

Those of us at Freeport who were working on the project knew all along, though, that building the 63-mile access road would be the toughest part of Phase IV. The road would originate at a tidewater depot on the Jaramaya River, an estuary of the Tipoeka, 11 miles upriver from a new coastal port to be built at Amamapare, 15 miles southeast of Timika. Heading north from the depot, the road would cut through 39 miles of mangrove swamp and rainforest jungle. At milepost 50—50 miles from the coast—it would ascend for 24 miles through steep mountain terrain to the mill site at milepost 74 in the canyon below the Ertsberg. The road would probably surpass aggregate road

construction for all of West Irian during the previous
decade.

There was no way to get around it, however. The mining and milling facilities could never have been constructed and supplied by helicopter. The huge ballmills
needed to grind the ore are 18 feet in diameter and weigh
over 20 tons. Tons of supplies would be needed every day
to keep the project functioning. A truck access road was
an absolute necessity.

Any lingering hopes I had that road construction actually might be easier than we thought it would were dispelled at the start. During February 1970, while we were
waiting for approval of our financing arrangements from
the German and Japanese agencies, I flew to West Irian
to see how preliminary work on the road was proceeding.
When I arrived at Amamapare, where we had located
what we called Camp One, I did not expect to see much
progress. The two bulldozers and four trucks I had ordered
had been on site for only a couple of months.

Yet I was totally unprepared for the scene that awaited
me. Camp One was in total confusion. Dozens of people
were sitting around waiting for orders. Equipment was
standing idle. Even though the on-site managers had spent
plenty of time scouting the swamps and making helicopter runs through the mountains to plot out the rough
outline of the access route, they apparently had been so
bewildered by the forbidding terrain that they had made
very few decisions. The trucks were supposed to be hauling gravel from riverbeds for the road north of the tidewater depot, but two of the trucks fell into rivers and the

failure to make a route decision immobilized the others. The four bulldozers, as planned, had been disassembled and flown by helicopter to the area near the old Base Camp where the town to house the work force was to be built. Two of the tractors were supposed to be pushing a pioneer or preliminary road through to the north while the other two were supposed to be heading south. But one of the northbound tractors was caught in a landslide after advancing up the mountain a few hundred yards. The driver escaped but the tractor today remains buried under 20 feet of landfill. The two southbound dozers actually progressed five miles toward the crest of a 9,000-foot ridge, but then one of them fell off a cliff into a 2,000-foot ravine—fortunately after its driver had jumped free—and was irretrievably lost. The other bulldozer became stranded at the edge of another cliff. Unable to figure out how to go any further, the driver abandoned the vehicle.

In short, the work that I had hoped would maintain the project's momentum had totally bogged down.

I immediately flew to San Francisco to confer with senior executives at Bechtel. It seemed clear that in their early planning they had underestimated what they would later say was the most difficult engineering assignment they had ever undertaken. Within two weeks, they dispatched to the jobsite two of their best and most experienced managers. The lost momentum was soon recaptured.

Our strategy for building the road was not like construction of a typical stateside highway. Merely pushing our way north from the tidewater depot would have taken too much time. Early on, we decided to use multiple points of attack, a series of construction bases spaced along the route from which road-building equipment would proceed both northward and southward.

The key to this strategy was the helicopter. On the Australian half of New Guinea, a gold-dredging venture known as Bulolo in 1939 became the first mining operation established using fixed-wing aircraft. The Ertsberg was certainly the first mining enterprise established with helicopters. At the peak of operations, 6 craft were in service manned by 11 pilots and supported by 10 mechanics.

The first task was construction of some 40 helipads along the road route. While the helicopter hovered just above treetop level, a man would be lowered on a cable rescue hoist 100 feet or so into the jungle. With a chain saw, he would clear a space about 20 feet in diameter. The fallen trunks would serve as the base of the pad. In the mountains, where large trees were usually not available, pads were built in flat areas with loads of two-inch planks. One platform used by a survey crew was perched precariously near the peak of a 10,000-foot mountain. Though stabilized by four braces, it could not be reinforced, sufficiently to support the weight of a helicopter. The pilot had to lower his craft gingerly onto the platform, being careful not to crush it in often strong winds, while his passengers hurried out with their equipment. One pilot, Dick LaFreniere, called landing on this pad his "most chilling experience."

Once, after a chopper had landed on a more secure pad near the crest of a ridge and unloaded some of its passengers, a strong wind gust flipped the craft over. With the pilot and a passenger still aboard, it fell into a 200-foot-deep gorge. The other passengers scrambled down toward the wreckage. In the cockpit, they found the pilot and the passenger apparently bleeding profusely. The blood, though, was actually red paint, a can of which had split open in the fall. Both men were shaken up but unhurt. We lost only one other helicopter, which had to make an emergency landing in a swamp. Everyone got out before it sank into the mud.

Once the pads were constructed, the choppers began unloading men, living facilities, supplies, and disassembled road-building equipment at the construction bases. Moving a disassembled D-6 Caterpillar tractor required as many as 36 helicopter trips. Often from two to five helicopters were required just to supply diesel fuel, which was airlifted in 2,250-pound loads of five 55-gallon drums. During one month—August 1971—the fleet of six choppers spent 636 hours delivering 1,495 loads of men and material averaging 2,250 pounds each for a total of 1,604 tons. The strain on the pilots was sometimes brutal. After several hours flying from sea level to 6,000 feet and back to sea level, they would often become so dizzy they would have to rest for a few hours before taking off again.

The easiest portion of the road, we had felt at first, would be between the swampland and the mountains, from milepost 39 to milepost 50. The region is covered by scraggly trees and a green undergrowth and from the air looks relatively flat and firm. Yet, as I had learned in my trip to the upper Carstenszweide, flatland in New Guinea that looks firm usually isn't. The dozers found that, somewhat like the swampland, the sandstone and quartzite bedrock is covered by an 8- to 12-foot layer of soft moss, peat, roots, and dead vegetation mixed with clay with the consistency of grease. While a man could usually walk on the surface, the dozers often sank into the muck and had to be pulled out by other dozers.

Before this section of road was started, there had been talk of building a cattle ranch in these flatlands. Some of the dozer drivers later remarked they would like to be around to see what happened when helicopters began unloading the first herd of cows.

The thickness of the muck varied considerably. To enable the dozers to carve out a route through the areas where the muck was shallowest, a foreman would regularly take the equivalent of soundings with a long jungle pole, as it was called. The pioneer road thus came to follow a very sinuous course. Normally twists in a pioneer road are straightened out later. In this case, though, we left the road right where it was. We thought building this 11-mile stretch would take five or six weeks. It ended up taking four months.

The greatest test of our bulldozer crew was in the mountains. While the best route through the mountains would seem to have been along canyon bottoms, most canyons end in jutting headwalls and are subject to frequent flooding and landslides. We had to use instead the tops of ridges and cuts into the sides of steep slopes. The route we chose led from milepost 50 on the flatlands up the often-precipitous south flank of Mount Hannekam along what we named Darnell Ridge after Bal Darnell, who had

headed our 1967 core-drilling program. In many places
the soft sedimentary rock on the ridge crest had been
honed by water erosion almost to a knife edge. The crest
was often only a couple of feet wide, so narrow that even
walking along it was hazardous. It had to be carefully
shaved by tiny D-4 tractors only slightly larger than a
riding lawnmower. This facilitated further leveling by the
larger D-6, which in turn prepared the way for the D-7,
D-8, and finally the giant D-9, a 25-ton monster with a
20-foot-wide blade known as "heavy iron." The tractors
often had to work in dense fog and rainfall of sometimes
more than a foot a day.

We hoped we could get by merely by slicing off the
top of Darnell Ridge, but the grade was so sharply inclined
in places—over 70 degrees—that we had to resort to an
elaborate series of switchbacks to reduce the grade of the
access road to no more than 25%; in other words, 25 feet of
vertical ascent for every 100 feet of horizontal advance,
about the maximum fully-loaded heavy-duty trucks can
manage. That is still more of a slope than the steepest
streets in San Francisco.

Grade reduction entailed moving an enormous amount
of earth: an average of 30,000 cubic yards a day. The total
for the whole ridge was 12 million tons, which is equivalent
to that dredged to build six miles of the Panama Canal.
At one point, we literally had to chop off the top half of
a mountain. One cut in the side of another mountain was
600 feet deep. Leveling Darnell Ridge took two and a
half months.

While the road to the north of Darnell Ridge required

less spectacular feats of earth-moving, it presented its own special set of adversities. While the Darnell Ridge rock is relatively solid, many sections of the northern road had to be carved out of the sides of mountains covered with slippery red shale and a surprisingly thick layer of dead vegetation. Making a cut in these slopes was like working with a pile of dry beach sand: for every cubic yard of material the dozers would push away, three or four yards of shale and vegetation would slide down into the cut from higher up on the mountain. Progress at times was only 10 feet a week.

The drivers of the Caterpillar dozers, who were led by a tough, 60-year-old veteran named Joe Singleton, are known as "cat skinners." The cat skinners were one of the project's elite subcultures of mainly free-lance individuals with highly specialized skills. Though of diverse nationalities, each subculture was tightly bound by pride and mutual respect. Their lives consisted of a continuous succession of brief tours of duty in all parts of the world.

The helicopter pilots, led by Don LeFreniere, who once flew for President Kennedy, were another subculture. And so were the tunnel drivers.

Originally, we didn't think we would need tunnel drivers. While we were fighting our way up Darnell Ridge, we organized several reconnaissance surveys to find a way to swing the road to the north along a cut in the slopes of Mount Hannekam several thousand feet below its summit. By the time we reached milepost 50 to 2,000 feet, though, our engineers and surveyors concluded that Hannekam's slope was too sheer. We would have to tunnel through the mountain. For the portal, we selected a point on Hannekam's face at an elevation of 5,800 feet that would later become milepost 58.

It would have been simpler to delay driving the tunnel until we had completed the road from milepost 50 to the tunnel entrance. The tunnel driving equipment could then have been brought in by truck. That, though, would have taken at least several weeks. Indeed, we hadn't even plotted out the road yet. In order to begin the tunnel immediately, we airlifted by helicopter six large truck-mounted pneumatic drills, several air compressors, tons of drill steel, ventilation tubes, camp buildings, repair shops, and thousands of curved steel tunnel supports, each weighing over a ton. All of this was deposited on a staging area near the portal created by leveling the top of an adjoining mountain.

Bechtel imported a hundred South Korean coal miners to serve as tunnel drivers and an expert group of expatriate supervisors. At the rate of a few dozen feet a day, the crew drilled and blasted a hole 20 feet in diameter that extended 3,627 feet through the middle of Hannekam.

When we started the tunnel, we not only didn't know just how its entrance would be reached by the access road, but we weren't sure where the breakthrough point on the other side of the mountain should be. We landed surveyors, often from helicopters perched precariously on ridge tops, to study the north face. Their calculations required us to change the direction of the tunnel 10 degrees to the east at its midpoint and increase the upward slope four degrees to make the breakout point compatible with what we had determined would be the route of the road to the north.

When the breakthrough came in September 1971, the road up Darnell Ridge and a major part of the road north of Hannekam was finished. Once the inside of the tunnel was ready for traffic, a long line of D-9's and other equipment too large to move by helicopter which had been waiting at the south portal made the first passage through the 3,627-foot hole. A few weeks earlier, a second tunnel beginning at milepost 70 and measuring 2,786 feet had been completed through the west shank of Mount Zaagkam, the only obstacle on the road to the mill site that we could not bulldoze away.

On Christmas Day 1971, the first convoy of trucks, carrying 500 tons of cargo, drove into the town site at milepost 68. We were now able to begin construction of the town. And most important, we were able to move our base of operations to the town site from milepost 39 on the flatlands.

The forward advance of the limit of truck travel permitted a rapid acceleration in the construction of the mountain facilities. In months, the town site was transformed from a rocky plain into probably the largest village in the southern part of the island, a community that will soon be able to house and provide the needs for 4,000 employees and their dependents.

In March 1973, President Suharto and a large delegation of Indonesian officials visited the project site for its official dedication. After his personal plane landed at the airstrip at Timika, he took the long and awesome trip up the road by jeep. At Tembagapura he was welcomed enthusiastically by the entire population. Speaking the next day at the dedication ceremony, he christened the town, and pressed a button which set the great ball mills near the Ertsberg rolling. At the same ceremonies, to our surprise, he announced that he had changed the name of the Indonesian half of New Guinea from Irian Barat, or West Irian, as it had been known, to Irian Jaya, or Great Irian. We had to send a bronze plaque referring to Irian Barat back to the United States to be recast.

The President's visit to the newly operating mine was a heartening demonstration of his interest in the Ertsberg project as the largest and most important development effort made in Indonesia's largest province. His interest has continued through the years.

While the houses in Tembagapura were being erected, trucks hauled up the road components of the mill, including the giant grinders and crushers that would reduce the ore chunks to a fine powder. Mill construction was completed on schedule. Not long after the mill began operating, though, we discovered a serious design defect. The mill had been modeled after other copper-ore processing plants in wide use throughout the world. Nearly all copper mines, though, are in rather arid climates. The weather around the Ertsberg is everything but arid. Instead of dust, tiny particles produced by the grinding and crushing process turned into mud, which stuck to and clogged up everything. "In less than six months," says a foreman, "the plant looked like it was 20 years old." A new "wet screening" process with water jets had to be installed to wash off the ore and restore the plant's efficiency. This proved to be the only design problem of any importance. The mill, which processes 375 tons of ore an hour, is now among the most efficient in the world, with a recovery rate of close to 97% of the copper in the ore it handles.

Perhaps the most challenging engineering feat was building a two-unit, 5,000-foot tramway system to move the ore from the mine to the mill 2,400 feet below. To hook up the first line, a helicopter took off from the canyon towing one end of 10,000 feet of quarter-inch rope and flew up to a tower near the mine. A man in the tower wrapped the rope around a drum and the helicopter flew back down to another tower in the canyon. The rope was made not of nylon but polypropelene. Nylon rope is somewhat elastic. If it had been used and then snagged and broken, the snap-back could have entangled the rope in the helicopter's blades and caused it to crash. Polypropelene rope stretches and stays stretched. Once the first line was in place, it was used to pull through the circuit ever-larger ropes and then cables. The final steel-track cables are three inches in diameter.

The original two ore tramways, designed by a Pittsburgh firm, were the longest of their kind ever built. In building the longest or biggest or heaviest industrial installation, one invariably runs into new and unpredictable problems that never bothered smaller facilities. The tramways were no exception. Shortly after it began moving ore, the cables started oscillating so violently that the 10-ton ore-carrying cars were sometimes derailed or flipped off onto the ground. We tried reducing the speed, cutting the load, lowering the center of gravity. Nothing seemed to work. Numerous consultants said they couldn't understand what was wrong. A world-wide canvass eventually produced a

Swiss mathematician who had extensively analyzed tramway oscillations. After elaborate calculations, he made some subtle modifications that greatly reduced the swings. "A tramway is like a violin," he said when he was finished. "It has to be tuned." We later went to a Swiss firm, Von Roll, to design a third, more sophisticated tramway with twice the capacity, which began operating in 1974. The three trams move in excess of 9,000 tons of ore a day.

The slurry pipeline is also the longest of its kind ever built. One trade magazine said it was "near the fringe of the state-of-the-art in slurry pipeline technology." Buried along the access road for most of its 69 miles, the 4¼-inch line would have to ascend and descend at grades that were regarded by some pipeline experts as unmanageable. Starting at an elevation of 9,200 feet, the pipeline would have to drop to 6,000 feet, climb to 9,400 feet, traverse a long stretch of ups and downs along a series of ridges, and then drop from 8,500 feet to 1,500 along Darnell Ridge in a span of only seven miles before gradually descending the last 50 miles to sea level.

Elaborate computer calculations and laboratory tests convinced us that such a pipeline could be made to work, and we went ahead and laid it. Copper concentrate in slurry form, though, can be extremely abrasive. When we pumped the slurry through the pipeline too fast or at too thin a consistency, it ate right through the pipeline walls on the steep sections. If we pumped it through too slowly or in too thick a consistency, it settled to the bottom and clogged the pipeline up, requiring replacement of the clogged sections. Only after many ruptures and abstruse mathematical calculations and laboratory tests did we get

the slurry right. The consistency of the concentrate had to
be kept between 64% and 67% and the speed of flow maintained at just over three miles an hour. It takes the concentrate about 22 hours to travel from one end of the line to
the other. When concentrate is not being pumped, water,
under carefully controlled pressure, must be kept moving
through the line.

No major technological impediments arose in constructing the port facility on the Tipoeka River. The job was just
miserable, nasty, and sticky. The port site at Amamapare
is in the midst of a section of the coastal mangrove swamp
that at high tide is covered by two to four feet of water.
Wading hip deep in smelly muck, chain-saw operators had
to cut their way through a thick tangle of roots and tree
trunks. The trees were left on the ground to serve as a
base. On top of the felled trees we pumped a 15-foot layer
of sand dredged from riverbeds and banks. To support the
buildings, we drove hundreds of 16-inch diameter steel
pipe pilings 80 feet into the ground, often right through
two-foot-thick mangrove logs. The result was a manmade
island on which are located living facilities for 300 employees and their dependents, a plant to filter and dry the
concentrate, and a pier from which the first copper concentrate shipment of 10,000 tons was made in December 1972
to waiting smelters in Japan.

The bugs have been gradually eliminated, but coping
with the demanding exigencies imposed by the project's
remote location and harsh weather continues to put a strain
on everybody. Our logistical managers try to keep in stock
100,000 items—everything from diesel fuel to aspirin, but

that is not easy. "When you're running a mine in Arizona and you need a part, you get on the phone and you can have it the next day," one foreman explains. "In Irian Jaya, it can take you four or five months to get something from the States." The logistics people have been able to cut that lead time by developing new, closer supply sources, mainly in Singapore and Australia. Instead of waiting nine months for truck parts from the U.S., we can often get them cus- tommade for us in four weeks by a machine shop in Singa- pore. Timber that used to be brought in from the West Coast of the U.S. is now purchased in New Zealand. Much of our food comes from Cairns, a small city on Australia's north coast, where Freeport relocated its main Australian headquarters after a hurricane leveled Darwin in 1974.

Perishable items are often brought in by plane on a space-available basis. A recent flight arrived with twenty- five cases of lettuce, six cases of mushrooms, three cases of broccoli, and six cases of strawberries. "It's amazing how fast the word gets around when we have fresh straw- berries," says logistics manager Orrin Main. "Those six cases disappeared in about fifteen minutes."

In emergencies we bring in heavy equipment by air. Once we had to fly in a two-ton diesel engine, but at $2 to $3 a pound for air freight, we do our best to move the heavy equipment by ship.

The supply system never works perfectly and people on the project learn to adjust to shortages. When the lettuce runs out, Tembagapura menus are routinely switched from salad to coleslaw made from locally purchased cabbage. Our machine shops have become adept in improvising repairs. Short-term fixes tide equipment over and keep it operating until the replacement parts arrive.

The omnipresent rain and fog that engulf the town and the mine most afternoons and many mornings wears everything out before its time—roads, vehicles, generators, people. "When the weather is clear and the sun is shining, everything breezes right along," says Doyle Dugan, general mine superintendent. "Equipment never seems to break down on a nice day. But when you're socked in, when it's rainy and cold and muddy and there's poor visibility, that's when you get all the problems."

Adapting to life in Tembagapura is no simple matter either. It is an isolated, constrained, and structured environment. Other than the town, there is only the road to the mine, the road to the coast, and menacing mountains on all sides. It is a closed, tight society with little privacy. Everybody works and plays together. The idea of "getting away from it all" doesn't have much meaning in Tembagapura. "No matter how carefully you try to tell people before they come what it's like, they are never quite prepared when they get here," says Dr. Cyril Swaine, a former Australian general practitioner and chief resident M.D., who has seen many people come and go during his eight years with Freeport Indonesia. "They have a hard time realizing there's only one store and you can't buy the same things that you can in downtown Tucson. If you're in the States or Australia and you get fed up, you can just get in your car and go somewhere. Here, you have to get along with what's available in town or you don't get along. By the end of the first three months, you can usually tell whether somebody's going to make it or not."

Some people make it and some don't. Some hate the remoteness, others find it relaxing. Some can't stand being out of touch, others like not having to worry about the problems of the rest of the world. Some get tired of the same old faces, others like the opportunity really to get to know their friends. One worker's wife left after two weeks. She couldn't stand the boredom, she said. It was too quiet. There was nothing to do. But another woman has been in town for several years and loves it. She entertains, reads, paints, sells crafts, gardens, works in the library, and jogs regularly. Her relatives back in the States complain that she hardly ever writes. She explains that she just doesn't seem to have enough time.

Even those who don't like Tembagapura very much are having an easier time of it, for the town has improved greatly over the years. The first structures had to be built as quickly as possible. There was no time for architectural frills or the niceties of sophisticated design. The town looked much like a group of Army barracks, row after row of gray buildings on mounds of gravel. Or, as a German friend of mine remarked when he first arrived, "Mein Gott, Stalag 17." But at the time, nobody seemed much concerned. Most of the town's residents were highly mobile transients, cat skinners, diamond drillers, and the like, who were used to getting by with far fewer amenities than were available in Tembagapura and who in any case did not much care where they lived because they never lived in any one place very long.

Once the construction phase was completed, miners, truck drivers, mill operators, and other more permanent employees became the town's predominant residents. Many of them planned to stay around a long time and brought their families. Gradually, the town's hard military look softened. The houses were painted different colors. Soil for gardens was brought in and bushes, trees, grass, and flowers planted. The bald construction scars on the nearby mountains slopes were reclaimed by jungle growth. New entertainment, recreational, and educational facilities were built. Tembagapura will never be just another pleasant suburban village, but it looks much more like one now than it once did.

Town life has changed in other ways. Most of the early construction workers were Western expatriates or "expats." Freeport has an obligation under its Contract of Work to employ as many Indonesian nationals as possible, and as the construction phase gave way to the operating phase, the town's population became increasingly Indonesian. The kinds of food on the supermarket shelves began changing: more kinds of rice, more spices, peanuts, coconuts, bananas, soybeans, and tumerics, a yellow root resembling a carrot. Indonesians began taking over many administrative and operating posts. Their style is different from that of expat managers, a little less structured, rigorous, and authoritarian.

Although working and living in the mountains in Irian Jaya is almost as difficult for Indonesians from Java, Sumatra, and other islands as it is for expatriates, Indonesians have become the backbone of the project's work force. They now constitute more than 95% of our employees, and their turnover rate averages about one-third that of the expatriates. Although little publicized, the fact that Freeport has facilitated the virtual takeover of the project by trained Indonesian nationals enhances the long-term benefits to Indonesia from the transfer of technology.

The melding of Indonesian and Western cultures has been enlightening for both, but not without some problems. Without realizing the potential for friction, for instance, early management let the employees' soccer teams organize themselves on ethnic and geographical bases. The emotions of the matches sometimes aggravated the rivalries of the various groups and caused rancorous incidents. Sports teams now are based on work departments, such as the mill, portsite, and townsite.

Basically, however, Indonesians have a great tolerance for diversity. For example, while the great majority of the population is Muslim, the rights of Christian and other minorities are carefully respected. Tembagapura is a diverse but representative community in this regard with active Muslim and Christian communities working and living together. Incidents have become rare.

Beyond its contractual obligations to employ Indonesians, Freeport Indonesia has felt a special obligation to hire people from Irian Jaya, particularly those from villages in the vicinity of the Ertsberg. By mid-1979, some 200 local Irianese were working on a more or less steady basis for the company. Most of their jobs are unskilled, ranging from road maintenance and sanitation to security and ship loading, but in increasing numbers, Irianese whose knowledge of machinery was once limited to stone axes have been taught to operate equipment such as fork lifts and dump trucks.

Employing local people has required an understanding
of the distinctive characteristics of their society and their
attitudes toward Western-style work. The impressions that
follow derive mainly from the observations of Frank Nelson, a Freeport geologist who between 1973 and 1977
worked closely with native employees. Those hired by
Freeport Indonesia come from several tribal groups and
in some respects are very different. Since they have had
little contact with the outside world, Amungme from villages in the mountains are very unsophisticated. By contrast, Irianese from such developed regions as Biak, Sorong,
and even from the heavily populated settlements around
Wissel Lakes, where there are many missions, are considerably more literate and worldly. Nelson found that
nearly all workers recruited from nearby villages tend
to display, in varying degrees, the same work-related
strengths and weaknesses, which Freeport has had to use,
adjust to, or attempt to change.

Especially important to the village lifestyle is clan
loyalty, which is stronger than tribal or especially national
loyalty. Clans are loosely organized extended-family
groups. Most of the men are closely related and many
are half brothers, for a man is permitted to have as many
wives as he can afford to keep.* Members of a clan seldom

---

* Missionaries in Irian Jaya have been trying to promote monogamy. The
most successful missionaries, though, are pragmatic and know there are limits
to the extent to which one can prudently tamper with local culture. One
missionary, whose flock was extensively polygamous, decided after a careful
examination of the Scriptures that the Bible did not really say that having
more than one wife was sinful. The Bible was, however, quite outspoken
against divorce. He concluded that while he would continue to urge bachelors
not to take more than one bride, it was not God's will that men who already
had several wives should be required to divorce the extra women.

extend aid or cooperate with members of other clans. Work performance improves greatly, Freeport has found, when native employees are grouped into small clan-related teams with their own areas of responsibility.

Even organized in clan-related groups, Irianese tend to have difficulty adjusting to alien Western work structures. Clan and village life is so simple that it does not require any of the involved and stratified levels of authority that characterize modern civilization in general and a large business organization in particular. A community typically will be led by a chief and a group of elders with charismatic personalities whose ideas best reflect the thinking of the people. Their leadership is informal and they operate mainly by consensus. A community will follow the advice of the elders but will switch its fealty if other men seem more worthy.

Local workers are confused when they must confront what they see as a complex hierarchy of chiefs with highly differentiated authority and responsibilities. As a result, they will often develop an exclusive relationship with one particular supervisor. Their resistance to cooperation with non-clan members does not seem to extend to foreigners. It is not unusual for a supervisor to find that a group of natives has formed a dependence on him for guidance in many areas beyond the job. In a sense, they make him their chief. In filling that role, the supervisor will be accorded a depth of loyalty from the local workers that his rank alone would never command.

Western job specialization in the beginning was very hard for the Irianese to comprehend. In a small village, everyone does much the same thing. No one person or clan specializes in raising pigs or sweet potatoes for exchange with someone else's axes or arrows. The local workers could not figure out how one man could be an expert in tractor-driving or radio operation while at the same time other employees had no idea how to operate these devices.

Many Irianese, though, have adapted to specialization with alacrity. They are often delighted to learn new skills that their fellow workers do not understand. They like being recognized as part of an elite group apart from the regular run of common laborers.

The rigorous schedules of Western work conflict drastically with local sensibilities. The Irianese sense of time is vague and unstructured. In village life, there is no special time to get up, perform a task, or go to sleep. There are not even any seasonal schedules to meet. Because Irian Jaya is so near the equator, the months are all pretty much the same and there is not much need to plant or harvest during a particular part of the year.

Nelson found that the best way to deal with the problem was to instruct local workers to be at the required place at least an hour ahead of time. Squatting on their heels, natives willingly will wait for hours without becoming bored or restless.

One significant exception to the unstructured sense of time is a local tribesman's periodic need to "go bush"—to take a month or so off a couple of times a year to return to his home village to take care of elderly parents and check the status of wives, pigs, and gardens. If he is not permitted to go bush when he feels he ought to, he will usually evince a loss in job interest, an increase in sick calls, and a general restlessness. Eventually the man will just leave anyway and may be afraid to return. The solution is regularly to give the local worker the requisite vacation time.

Counterbalancing the adjustment difficulties, Nelson found, was a commendable industriousness among many Irianese, often greater than that of Western workers. They are used to heavy, hard, uncomfortable work and admire physical toughness and courage. Most have a high mechanical aptitude. They have no preconceived notions about machinery or working, and come to the job with an open mind and a desire to learn. They are surprisingly quick in adapting to new conditions and situations. And they tend to maintain a fine morale and sense of humor unless they are yelled at, in which case they assume a dull, stupid pose, or laughed at, in which case they can become very upset and even violent. As long as they are carefully selected, trained, supervised, and assigned to the appropriate tasks, the local Irianese, Freeport Indonesia has found, constitute a promising element of the project's work force. Of course, many years of education and living and working in a modern environment will be required for the local Irianese to reach the competency levels of our other Indonesian employees.

As a twentieth-century industrial installation implanted in the middle of a Stone Age society, the Ertsberg project has had a broad impact on the life of the natives in the area. The full consequences will not be known for many years, but some changes are already apparent.

Native employment is accelerating the gradual replacement of the traditional barter economy with a more formal money economy. Workers are being paid in Indonesian rupiahs, which they use to buy food and other goods in the store in Tembagapura. Other local villagers are earning money through the sale of vegetables and handmade artifacts such as spears and necklaces.

Some transactions, though, are still not amenable to a money economy. Not long ago, a girl of perhaps 14 from a mountain village broke her leg when she was accidentally hit by a truck in Tembagapura. The leg was promptly mended at the town's hospital. It developed, though, that the girl had recently been purchased by a local chief as his future bride. Taking the position that Freeport Indonesia had damaged his purchase, he indicated he was considering exacting a couple of lives in Tembagapura as reparation. A Freeport official eventually negotiated the chief's drastic terms down to one large pig. The best the official could come up with right away, though, was two medium-sized pigs. The chief agreed to the substitute, which was brought in by airlift, and took his bartered bride and the pigs home.

The notion that one who suffers an injury is entitled to reparation or revenge is a facet of a broader and very important tenet of Irianese society that has been affected by Freeport's presence: equivalence. The equal exchange of goods and services is regarded as the main underpinning of social relationships. One observer, Peter Lawrence in *Road Belong Cargo* (Humanities Press, 1967), described the principle this way in discussing the Melanesians' material value system:

It was important that the exchange of goods and services be equivalent. This was based on their lack of

specialization which resulted in everyone having resources which alone made the principle of equivalence
a reality. Where there was no exchange of goods and
services there could be no sense of friendship, mutual
obligation and value, but only suspicion, hostility,
and the risk of warfare. Nor could they conceive of a
friendship that did not result in approximate economic
equality.

Due to equal access to economic resources and the pau-
city of material wealth generally, private property is less
important to Irianese society than to Western society.
Individual possessions are limited mainly to clothes, shelter, and tools. Food and land are widely shared. A clan will
acquire a right to a plot of land it occupies or cultivates,
but actual ownership is considered to be communal.

The unavoidable wealth disparity between local villagers who work for and trade with Freeport and those who
don't has sometimes led to tension and resentment. A
worker will usually share with others in his clan and often
will support several people on his salary. Many workers
seem pleased with this arrangement, for one measure of
status within a clan or village is the amount of goods one
shares with others. A man wins respect not by the number
of pigs he owns but by the number he donates to village-
wide pig feasts. Some workers, though, have become less
inclined to share and more caught up with Western notions
of private property. A few have opened their own accounts
at the Tembagapura bank. In some villages, the workers
are coming to constitute an economic hierarchy that is be-

ginning to affect traditional social structures such as the authority of elders and chiefs.

The severity of the impact of the Ertsberg development varies with proximity to the project and previous contact with foreigners. The effect on Biak, Sorong, and the Wissel Lakes settlements is probably very small. It has been somewhat greater on villages like Beoga, a small, remote community of several hundred in a valley on the northern slope of the Jayawijaya (formerly Carstensz) range that is several days journey from Tembagapura. According to Don Gibbons, an American missionary who has lived with his wife in Beoga for 20 years, the men from the village who work for Freeport Indonesia have become more aggressive and status conscious than others in Beoga, where life is very placid, leisurely, and communal. Gibbons adds, though, that the town's social structure remains generally undisturbed and its economy has benefited. Beoga gets $150 from the 1,500 pounds of vegetables it sells each week to Freeport. The Freeport employees from Beoga send most of their money back home and one recently donated $100 toward the cost of a new church roof. "The benefits," concludes Gibbons, "outweigh the disadvantages."

Nearer Tembagapura, Gibbons feels, "there has been a significant disruption in the native culture." About 200 men from Wa, Tsinga, and other primitive mountain villagers work at the mine, mill, and town. About 20 coastal villagers work at the port facility. Two squatter settlements have sprung up on the outskirts of Tembagapura, where about 100 native employees, their families, and various others reside in makeshift metal and wood shacks. Most of these people have given up their traditional clan and tribal roles and have developed dependency relationships with Freeport. The relationships include begging and occasional theft, although there is probably less theft than in a typical large U.S. urban area. Theft is usually rationalized by the idea of equivalence: Since Freeport occupies what the local Irianese consider to be their land, the goods they take are deemed proper sharing of Freeport's possessions.

Availability of employment, opportunities for pilferage, and simple curiosity has caused migration into the project area of some tribal people from distant villages. This in turn has caused hostilities between the migrants and the locals. When the economic opportunities have not proved as attractive as was rumored, the migrants have become angry.

Upset by the disruptions in village life, local chiefs at times have complained to Freeport. As they see it, Freeport has violated the precept of equivalence. The company has wrongfully appropriated their land without adequate compensation. Though Freeport employs many locals, the chiefs complain they have lesser jobs and receive lower salaries than workers from other parts of Irian Jaya. The chiefs have a hard time understanding that the men from more advanced sections of the country are more skilled and better able to adapt to industrial work. To the chiefs, all men are equal and should receive equivalent wages and occupy equivalent positions.

The chiefs have occasionally showed their displeasure by putting hex sticks around Freeport's facilities. In 1973, after some villagers from Tsinga had staked a small exploration camp in the mountains east of the Ertsberg, Frank Nelson, Dr. Cyril Swaine, and John Ellenberger, a local missionary who served as interpreter, flew to Tsinga to confer with local leaders. They were greeted by close to a hundred angry villagers wearing battle dress—painted faces and feathers—and heavily armed with spears and bows and arrows. For several hours, they yelled at the Westerners, shook their weapons menacingly, and even seemed to threaten to eat the visitors.

The town chief, a wizened old man, finally stood up and commanded everyone to be silent. He was tired of all the arguing, he said. "The only thing we want that you white people have to give us is the gospel of Jesus Christ. Now stop talking about that rock and teach some of the gospel before you go." Whereupon Ellenberger, who can be a spellbinding orator, delivered a brief sermon and said a prayer. The tension was broken. Women emerged from huts with gifts of bananas. Amid laughter and joking, the relieved visitors boarded the helicopter and took off for home.

Underlying the feelings of these tribal people, and making the situation especially hard for Freeport to deal with, is cargo cultism, which is pervasive in Melanesia. The Ertsberg project has stimulated local fascination with modern material goods and spawned numerous cargo myths. A man from a village to the west arrived in a valley near Tsinga one day a few years ago claiming to be a prophet. He said that Freeport had a key to a *pabrik* or factory inside a mountain from which the company had obtained its machines and other items. The key was the tooth of a *naik-ma nung op* or "big-tooth rodent." If the villagers brought him the right tooth, the man said, he would open the *pabrik* and they would have all the material goods they had ever seen the foreigners possess. Though the prophet when provided with the requisite tooth proved unable to unlock the mountain, he was venerated by nearly the entire population of the valley for several months.

Many local villagers are certain that Freeport possesses such a magic key. They are not satisfied with the company's occasional gifts of material goods. Many are not very interested in the company's offers to train them for jobs or teach them agricultural skills. The notion of equivalence requires that the local people immediately receive the same material affluence as the white men who have taken over their land. Some village chiefs have demanded that Freeport give them trucks and build a road from their villages to the mountain where the cargo comes from so they can obtain for themselves their rightful share. Some have simply asked for a road. Once the road is built, they feel, cargo will magically begin to flow into their villages.

The local villagers' cargo beliefs made them especially susceptible a few years ago to the blandishments of individuals associated with a dissident separatist movement. This small and rather ragtag band has sporadically resisted Indonesian rule over Irian Jaya since 1963. It has tried on occasion to ambush Indonesian army patrols and to hold hostage and assassinate district officials. The campaign intensified after Papua New Guinea obtained its independence in 1975 from Australia. Indonesian troops parachuted into the Baliem Valley the following year to quell an uprising which involved some bloodshed.

In 1977, a few rebels from areas near the Papua New Guinea border tried to recruit Irianese in some of the villages near Tembagapura for a series of attacks against Freeport Indonesia's installations. Their apparent interest was not in Freeport Indonesia as a company but as a highly visible symbol of the Indonesian government. They told the villagers that the Indonesian government was blocking the flow of Freeport's cargo to the local people and that when the government was driven out the villagers would become as affluent as the foreigners.

Many Irianese, including a handful on Freeport Indonesia's payroll, were persuaded and left their villages to join the rebels. Over a period of six weeks, they cut the slurry pipeline in a number of places, ruptured fuel and power lines, and burned an explosives magazine. The Indonesian government, which has always given strong support to the Ertsberg project, responded promptly by flying in troops to safeguard the facilities. The company occasionally had to suspend operations to make repairs.

Life in the jungle with the dissidents proved a lot harder for the new recruits than living at home. Many became sick and died. Others became disillusioned when the cargo bounty was not forthcoming. A few months later, they started trickling back home. Most of the Freeport Indonesia employees who had joined the rebel group were apprehended. No more attacks have occurred since September 1977 and rebel influence in the project area appears to have disappeared.

Freeport has been doing what it can to improve its relations with nearby villages. Its philosophy has been to resist demands for cargo and avoid charitable, paternalistic activities that might foster excessive dependence. Despite their lack of skills and their trouble adjusting to modern work organizations, as many local people as possible have been hired. Freeport continues to increase its purchases of local produce and artifacts.

The Tembagapura hospital regularly treats local villagers. Women whose husbands once forbade them to visit the hospital now come to have babies delivered. One requiring a Caesarean section was flown to a hospital in Cairns. Not long ago the chief of a village near Wa was carried in with 11 arrow wounds suffered during one of the frequent tribal wars. Though gravely ill, he was treated and recovered. When he went home a few weeks later, he told the town's general manager that it was a good thing for Freeport that he had survived. If he hadn't, he said, there would have been a lot of trouble.

Freeport Indonesia is careful to avoid getting enmeshed in local disputes. One afternoon, the chief of Wa came into town complaining that another Wa resident, who was a Freeport employee, had absconded with one of his daughters without paying the bride price. The chief's men had shot arrows at the man and the woman as they fled into the jungle, but all had missed. The chief wanted a helicopter to press the pursuit. The company politely declined.

Sometimes Freeport gets involved against its will. Once residents of Wa and two nearby villages took on members of the Moni tribe, which lives to the west. Monis usually acquit themselves well in tribal battles, but this time they were nearly wiped out because the Wa warriors had come up with a surprise technological advance in the art of bow-and-arrow combat: shields made from plywood and suspended from the neck by electrical wire "borrowed" from Tembagapura.

After many attempts to improve the living conditions of mountain villages, Freeport officials have gradually concluded that life in the highlands is inherently just too formidable. According to Wyn Coates, assistant to the president of Freeport Indonesia and a former American foreign service officer fluent in Indonesian, "The best that is possible is marginal subsistence living. If they stay in the mountains, nothing can change."

The provincial government and Freeport are now building a village in the flatlands where they hope the mountain people living near Tembagapura will resettle. The new town will have 300 houses, a clinic, a school, and a store stocked by Freeport. The central government in cooperation with the local administration and the company will assist residents to raise crops and domestic animals that can supplement pigs as a protein source. But Coates concedes it is far from certain that very many of the mountain tribesmen can be persuaded to move. When several hundred resettled in a village started by the Dutch in the coastal plains during the 1960s, many caught malaria and other diseases and perished. Many others went back to the mountains and the project was very slow to develop, although it now has more than a thousand inhabitants.

Coates says, further, that the company is unsure of the best ways to teach the Irianese to help themselves and improve their welfare. The company built a latrine and an incinerator in one of the squatter villages just outside Tembagapura, but it was never able to convince the villagers to use them. When it tried to get the villagers to raise chickens and rabbits, it found they didn't want to have anything to do with animals that, unlike pigs, cannot take care of themselves. Freeport hopes to experiment with goats, which are more self-sufficient. "You have to understand that we're working on the basis of very little information about the Amungme/Damal culture, which nobody has ever studied in depth," Coates stresses. "What motivates them? What is valuable to them? What do they consider as incentives? How do they relate to their peers and the outside world? We just don't know very many of the answers."

Del Flint once suggested, only half in jest, that maybe we should have built a huge wall around the Ertsberg project so that it would not have any ill effects on the local people. That, of course, would have been impossible. "There's no way you can cut yourself off," says Wyn Coates. "We've had to come to grips with the fact that we're part of the local cultural scene whether we like it or not."

Freeport Indonesia's activities are just one facet of the broad encroachment of modern civilization on virtually all of the primitive societies of Irian Jaya and, indeed, the world. The consequences are often profound and disturbing. Shortly before he disappeared off the coast of New Guinea in 1961, Michael Rockefeller wrote to his parents about his impressions of the Asmat, a remote tribe living a hundred miles down the south coast from Kokonao in the vicinity of the town of Agats:

The Asmat is filled with a kind of tragedy. For many of the villages have reached that point where they are beginning to doubt their own culture and crave things Western. There is everywhere a depressing respect for the white man's shirt and pants, no matter how tattered and dirty, even though these doubtful symbols of another world seem to hide a proud form and replace a far finer . . . form of dress . . . The West thinks in terms of bringing advance and opportunity to such a place. In actuality we bring a cultural bankruptcy that will last for many years . . . There are no minerals; and not a single cash crop that will grow successfully. Nonetheless, the Asmat like every other corner of the world is being sucked into a world economy and a world culture which insists on economic plenty as a primary ideal.

Even if it were possible, though, there are many grounds for questioning to what extent the local villagers' way of living *should* be preserved like that of some endangered species in a wildlife refuge. "Their infant mortality is close to 50%," says Wyn Coates. "Their food supply is terrible and there is wide malnutrition. Their average life-span is 35. They are 99% illiterate. Are those the sorts of things you think ought to be preserved?" Many aspects of life in local villages, certainly, are valuable. However, Coates says, "It is hard to change certain things while preserving other things. When you change one thing everything else changes also." One cannot raise a village's standard of living by improving nutrition, increasing literacy, teaching residents specialized skills, and helping them organize productive enterprises without upsetting the village's traditional social and economic structures. "Modernization is not beautiful to watch up close," Coates sums up. "It is an unsettling, disruptive process." But for the tribal people of the Irian Jaya mountains, the process, sooner or later, is inevitable.

Ever since we paddled upriver in 1960, the Ertsberg project has encountered and overcome a bewildering variety of geological, technological, political, and economic obstacles. In the early days, many of them could well have killed the project. Some came very close to doing just that. When I retired in 1974, we still had troubles. We were still trying to figure out what to do about the hex sticks and angry tribal chiefs. But I had lost at last the fear I had had on and off for so many years that the world would never have the benefit of the Ertsberg's riches. The project had achieved such a momentum that its success was assured.

Between December 1972, when the first token shipment of 10,000 tons of copper concentrate left for Japan, and the end of 1980, the Ertsberg had produced nearly 23 million short tons of ore from which we had obtained 521,000 tons of copper, 6 million ounces of silver, and 463,000 ounces of gold. These metals had a gross sales value in excess of $772 million.

Profits, though, were another matter. In its early years of production, the Ertsberg fully justified the grim predictions of its detractors. During the mid-seventies, in particular, when copper prices were unusually depressed, Freeport Indonesia operated deeply in the red. Without drastic economies and enlightened forebearance by Indonesian tax authorities, we might have had to suspend operations. As has been mentioned, dividends from Freeport Indonesia to the parent company were infrequent and small and the meager return the parent company realized in its large investment was far from commensurate with the risks.

During 1979, the Ertsberg at last began to produce the sort of profitability that the project's supporters had envisioned during the long years of exploration and development. World copper supplies became very tight and prices for a time soared past $1 a pound, nearly double the level a few years earlier. Gold and silver prices hit record highs. Freeport Indonesia in 1979 contributed $30 million to the parent company's earnings. In 1980 its contribution of earnings to the parent company amounted to nearly $50 million. The copper market, it should be stressed, is notoriously cyclical. While many observers are predicting a tight market and high prices well into the 1980's, there is no guarantee that the present level of profitability will be maintained. Nonetheless, those of us who fought so hard for the Ertsberg project are justified, I think, in feeling vindicated.

That feeling comes at a time when the Ertsberg is close to depletion. Only a few million tons of ore remain and that will be gone by 1984. The onetime fog-shrouded monolith soaring up over 500 feet is now a huge but much less enthralling open pit several hundred feet deep. Some of the people who worked with me during the exploration phase have become rather nostalgic. Not long ago, Pete Petersen was looking at slides of the pit. "It never really hit me that the Ertsberg would disappear," he said. "It hurt. That rock was a symbol."

I always recognized it would disappear, and I have never felt nostalgia or mixed feelings. My objective was always to get a commercial operation going, to get that ore body mined. In the early days, one of the Australians suggested the Ertsberg be renamed Wilson's Knob. "Hell, don't name it for me," I replied, "because it isn't going to last very long."

Petersen and others who feel nostalgic can be consoled by the presence of at least one other ore body to take the Ertsberg's place. In 1975 we began exploration of the Flint Extension, as we called the area of mineralization discovered by Del Flint, in 1968. Though less than a mile to the east of the Ertsberg, the new ore body, renamed Ertsberg East, was in such treacherous terrain that test drilling had to be supported entirely by helicopter. Some of the drill stations were at elevations of over 13,500 feet. Many were mere toeholds in sheer cliffs. To reach two of these stations from the drilling camp, drillers had to climb a series of ladders 500 feet high.

While initial results showed high-grade copper ore, sufficiently powerful drills could not be placed in enough locations to prove up reserves required to justify production. Since we couldn't get at it from the top, we decided to reach the deposit by going underground. Starting near the open pit, we drove a 4,000-foot adit through the intervening igneous rock to intersect the ore body. Freeport Indonesia now estimates that the deposit contains more than 45 million tons of recoverable high-grade ore. To develop this new underground mine, commitments for up to $101.5 million were obtained from a group of United States and Canadian lenders. At the end of 1980 the mine development program was 90% complete and substantially under budget. Ore from the underground block caving operation will be available in the first quarter of 1981. Ultimately, Ertsberg East could produce as much copper as three to five Ertsbergs. And there may be still other ore bodies nearby.

Nearly all of the original Ertsberg may have been ground up, slurried down the mountains, and shipped to Japanese and German smelters, but by my way of thinking, that majestic outcrop of copper endures in spirit if not in fact. It is still a towering symbol for us all.

Appendix Notes

---◆---

Pre-1960 Expeditions to the New Guinea Highlands

1. In preparation for my trip, I conducted extensive research into the handful of earlier expeditions into the highlands of New Guinea. Most, I found, had failed because of a poor choice of routes. The earliest recorded venture was by two captains of the Dutch army, one with the implausible name of Posthumus Meyers and the other known only as De Rochemont. From 1903 to 1905, they explored many rivers along the south coast by boat and made several excursions into the foothills, but never found an access route to the Carstensz Top. After his 1907 expedition to climb 15,518-foot-high Mount Wilhelmina, 60 miles west of the Carstensz Top, foundered due to lack of porters and provisions, Dr. H. A. Lorentz, a Dutch explorer, returned two years later with a mammoth retinue of 247 people. It included 179 Dyak bearers from North Borneo and an impressive military contingent supported by 20 convicts, it

being Dutch practice in those days to use convicts from Java and Sumatra as porters for military forces. Ten months later, Lorentz reached Wilhelmina's lower slopes, but exhausted and nearly out of food, he never made it to the top. On the two-month journey back, Lorentz was laid up for three weeks with broken ribs and severe contusions suffered in a fall. He later estimated that the Carstensz Top was 18,000 feet high and would never be climbed.

An approach to the Carstensz Top via the Mimika River in 1910 by William Goodfellow, a British naturalist, had advanced only to the foothills; a tenth of his party had perished from malaria and other misfortunes, and Goodfellow himself had become so seriously ill that he had to return to England. His assistant, Captain C. G. Rawling, took charge and was joined by A. F. R. Wollaston, a noted mountain climber who was later the first European to explore the lower slopes of Everest. But after 15 months, they were only able to reach an unclimbable escarpment that Rawling estimated was 10,000 feet high on the south face of what was then known as Mount Leonard Darwin, a peak just west of the Carstensz Top. (Rawling later renamed it Mount Idenburg, in honor of the Dutch governor of the territory.)

Wollaston returned in 1912 with a 226-man force and provisions that included 17 tons of rice, 100 pounds of tea, 1,200 pounds of sugar, and 25,000 rounds of ammunition. Following the Otowka River up from the coast, he became the first white man to meet the pygmoid Irianese who lived in the highlands, but four and a half months later, at an elevation of 14,866 feet on the southeast flank of the

Carstensz Top, he was forced to turn back by a huge ice fall at the foot of a high, nearly vertical limestone cliff.

Mount Wilhelmina was finally scaled in 1913 by a Dutch government expedition, but after World War II, the most intrepid European mountain climbers turned their attention to the more challenging Himalayas. The Carstensz Top remained unclimbed until the arrival of Jean Jacques Dozy.

The Strange Megapode Eggs

2. More than a year later, after my 1960 expedition, I described the unusual eggshells I saw during my rest stop on June 3rd to my good friend Tom Gilliard, Associate Curator of Birds at the American Museum of Natural History in New York. Tom had made many trips to the interior of Australian New Guinea, as well as to other isolated parts of the world. In addition to collecting and naming dozens of new bird species and subspecies, he is recognized as one of the world's foremost authorities on birds of paradise and unusual bird forms found only in this section of Western New Guinea. According to Tom, the eggs were from a family of birds variously called "brush turkeys," "mound builders," or "thermometer birds." More accurately, they are known as megapodes, or big feet. They are the only known order of birds that utilize heat other than their body for the incubation of their eggs. Like young turtles, alligators, and snakes, young megapodes are hatched beneath earth, sand, forest litter, and even volcanic ash in areas where volcanism is still active.

The bird that laid the large eggs we saw was probably a variety of megapode known as telegalla whose mounds are often found on the steep slopes above the coastal low-lands of New Guinea. The telegalla's habits are described in Tom's magnificent book, *Living Birds of the World.*

Pairs or groups of birds rake together the debris that lies under the tangled brush. Each bird moves slowly backward, raking with its huge nails and feeding on grubs and insects thus exposed. It stands on one foot and rakes powerfully with the other, sending dead leaves, sticks and all manner of litter cascading back-ward toward an ever-accumulating heap. Sometimes the mounds reach a height of 15 feet, but usually they are from 5 to 7 feet high, about 20 feet in diameter and roughly conical.

After completion and maturation the compost-like mound warms up to 95° or 96° F. It is then that the heat-sensitive female excavates high on the side of the heap a hole several feet deep and slanted steeply in-ward. In this she lays from five to eight large, white, oval, thin-shelled eggs, drilling a new hole for each egg.

The period of incubation, up to 63 days, is the long-est known in birds and contrasts strikingly with 11 to 12 days for the cowbird, and 21 days for a domestic fowl. During this period each pair of megapodes at-tends the mound which must be aerated for tempera-ture control!

When the young bird finally emerges from its incubator, its feathers are free of their waxy sheathing and the plumage is so well developed that if necessary it can spring into the air and fly, though it is still relatively small and quaillike in appearance. It is fully able to and frequently does take care of itself from the moment of hatching.

If all these interesting facts had been known to me at the time, I am sure that I would have been tempted to seek out the mound from which these eggs had been obtained.

Wollaston's Ridge-Crest Encounter

3.   After returning to the U.S., I found the following in Wollaston's account of his 1912 expedition:

On the next day our track lay along a knife-edge ridge between the Utakwa and one of its principal tributaries, the Nusala-marong, several thousand feet below us, and there it appeared that we passed from the region of one tribe into that of another. After crossing a rough wooden fence set across the ridge, we came upon an open level space about the size of a lawn-tennis court, where we found a crowd of 60 or 70 people of both sexes and all ages, who set up the most astonishing barking noise when we appeared. Many of them pranced or danced up and down the platform and others crowded round and shook hands, or rather pulled knuckles with us.

After a time we were told to stand with our party at one end of the platform while the people who belonged to the country beyond stood at the other end, an open space left between us. Then a man, who appeared to have some authority, ordered silence and began a long harangue; in one hand he held a stone axe and in the other two white leaves, with both of which he gesticulated freely in our direction. Toward the end of his speech a lean white pig was brought from the back of the crowd, and I was instructed to go forward and receive it. Fortunately, the man of authority presented me with a small boy and girl as guardians of the animal, and we agreed to fetch it on our return, but, of course, we never saw it again. We gave them small presents and were then allowed to continue our way along the ridge, which we followed for two days' further march.

The English and Dutch explorers who first penetrated the south coast of New Guinea tried to set down the names of rivers as they heard them from the native people. In this account I have used those accepted by Father Koot. Hence, for Wollaston's Utakwa read Otokwa (Tsing in its headwaters), and for its tributary, Nosolonogong. While there are many dialects among the numerous tribes and family groups, a common word is *gong* (stream or tributary to a larger river). Hence: Kelogong, Keemogong, Nosolonogong, etc.) Interestingly, *gong* has a similar meaning among Australian aborigines.

Vegetative Gigantism at High Elevations

4. Paul Zahl, a botanist, once wrote an article called "Mountains of the Moon," describing curious plants growing at elevations of 10,000 to 13,000 feet on the slopes of the snow-covered Ruwenzori mountains in central Africa. Zahl described several varieties that had attained outlandish sizes: groundsels as tall as telephone poles, and heath trees, the well-known heather of Scotland, reaching heights of 40 feet despite a smothering mantle of moss clinging to every branch. Botanists have not been able to explain such gigantism. Why do little plants in ordinary environments try to become trees in such remote places as the Carstenszweide and the upper slopes of the Ruwenzori? One theory is that being at such high elevations, they are subject to more intense cosmic rays. Zahl's initial thoughts while observing the vegetation were the same as mine in New Guinea. "As we hacked our way through," he wrote at one point, "I would not have been surprised to meet a browsing dinosaur or a soaring pterodactyl."

INDEX

237

Drilling for Ertsberg core samples,
149, 150, 165, 166, 169–170,
174, 177, 180. *See also* Development of Ertsberg, Phase II
Drums, 144
Dugan, Doyle, 205
Dugout-canoes, 24–25, 30, 37, 38,
45–46, 140, 141, 142
Duke, Bob, 158, 160
Dutch colonization of New Guinea,
4 and *n.*, 30, 151–153

East Borneo Company, 13–14, 15*n.*,
105, 114
Eastern New Guinea. *See* Papua New
Guinea
Ecuador, 79
Electrical conductivity of copper, 8
Ellenberger, John, 217
Engineering studies, for Ertsberg
development, 171, 178, 180, 182–
184, 186, 187
Equipment:
Ertsberg expedition, 33, 35, 40–41,
51–52, 67, 109, 114, 114*n.*–
115*n.*, 123, 137 (*See* also Air-
dropped supplies; Radio equipment)
sources of, for Ertsberg development, 203–205
Equivalence, importance in Irianese
social relationships, 213–214,
216, 218
Erosion:
due to rainfall, 111
glacial, 112, 113
Ertsberg, 8, 104–105
aerial survey of, 27–28, 92, 197
air approach to, 25–26, 33, 150,
152, 165–169, 179, 180
climate near, and effect on Carstensz glacier, 106 and *n.*
copper discovery at, 12–13, 14
development of. *See* Development
of Ertsberg
discovery and first climbing of, by
Dozy and Colijn, 6, 10–13, 119
formation of, from glacial erosion,
112, 113
land approach to, 26, 28–29, 45, 46
largest known above-ground copper
deposit, 7, 112, 148
mineralization of, 111, 148, 149,
174, 175
ore deposits of. *See* Ore deposits,
Ertsberg
See also Ertsberg expedition
Ertsberg East, 117–118
ore deposits of, 226–227

Ertsberg expedition:
air-dropped supplies for, 41–42, 43,
81, 82, 107, 122, 135–136
end of, 146
food supplies, 65, 66, 82, 96, 127,
135–136
funding for, 15 and *n.*
health problems, 56, 61–62, 64, 65,
66, 68–69, 74–75, 99, 118, 125,
132–133, 143
Irianese supply looting of, 129–130
ore sampling and surveying, 110–
123, 149
radio equipment and communication, 33, 35, 41, 67–68, 81–82,
83, 109, 127
route selection, 25–29, 86
traveling difficulties of, 56, 58–59,
60–61, 62–63, 77–78, 80, 86–87,
88–90, 103–104, 131–132, 134,
138–140
tribesmen as load carriers for, 32,
35, 37–38, 46, 51, 52–53, 64–65,
66, 75, 81, 86, 95, 115, 121, 122,
126, 127, 130, 135–137, 143
Export-Import Bank, U.S., 184, 185

Faults, 111
Feldman, Dr., 25, 28, 145
Fitzgerald, Ted, 164–165, 166
Flint, Delos, 26, 33, 34, 54–55, 67, 68,
87, 88, 90, 104, 108, 109, 124–
125, 131, 139, 146, 169, 177,
222, 226
Flint Extension, 226
Food supplies, for Ertsberg expedition, 82, 96, 127, 135–136
shortage of, 65, 66
Footwear, for hiking, 74, 75
Fossils, 111
Freeport Indonesia, Incorporated,
15*n.*, 26, 148, 149, 152, 153, 170,
181, 183, 184, 186, 188, 207
employment of local Irianese, and
problems with, 208–222
hesitancy on Ertsberg development,
175–176, 178, 186
negotiations with Indonesian government, for development of
Ertsberg, 154–163
Freeport Minerals Company, 6, 9,
15 and *n.*, 160, 185–186
Freeport Sulphur (mining company),
6, 185
Funeral pyres, 85

Garnaut, Ross, 5
Germany, 184, 185, 186, 188, 191
Gibbons, Don, 215

Born in York, Maine in 1910, Forbes Wilson was educated at Worcester Academy and Yale University. At Yale he won the Samuel Penfield Prize in Mineralogy and the Silliman Geology Award, graduating with a B.S. degree in mining geology in 1931.

Wilson's career as a mining engineer led him first, for two years, to Chile with the Braden Copper Company, and then, for nine years, to Colombia, as manager of a gold mine. His association with Freeport dates from 1942, when he was assigned to the Nicaro Nickel Company in Cuba. As general manager of that company he returned to Freeport's New York headquarters in 1947, and shortly thereafter was placed in charge of a world-wide mineral exploration program that involved extensive travel in Canada, Ireland, Tunisia, Spain, Greece, Turkey, Australia and Indonesia. In 1966 he became president of Freeport Indonesia, Inc., a post he held until his retirement in March 1974. He remains a director of that company.

Early in 1977 the Society of Mining Engineers gave Wilson the Daniel C. Jackling Award, "For his vision, determination, dedication and leadership in the conquest of the remote and rugged Ertsberg and of the technical and human barriers to its development." Later that year, the Society named Wilson a Distinguished Member in recognition of and appreciation for his outstanding service to the mining industry. In 1980, he received the coveted Legion of Honor Award from the American Institute of Mining Metallurgical and Petroleum Engineers.

The son of a professional golfer, Forbes Wilson has himself excelled at that sport, winning a number of Maine state championships. In 1980 he achieved an objective of all senior golfers, shooting a 69, one shot under his age, on a par 72 golf course. He and his wife have five daughters and three grandchildren. They live in York, Maine.

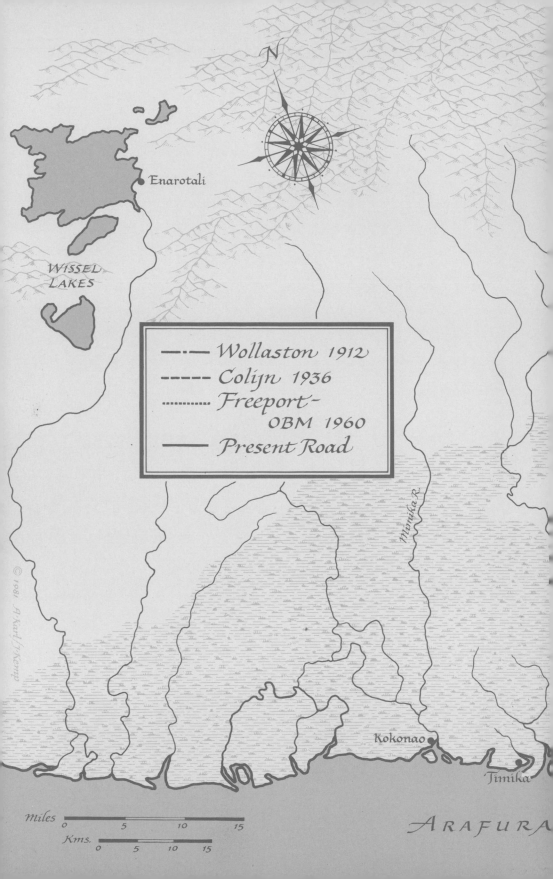

N

Enarotali

WISSEL
LAKES

--·-- Wollaston 1912
----- Colijn 1936
············ Freeport-
    OBM 1960
—— Present Road

Manika R.

© 1981 A Karl/J Kemp

Kokonao

Timika

Miles
0   5   10   15

Kms.
0   5   10   15

ARAFURA